Palgrave Advances in Luxury

Series Editor
Paurav Shukla
Southampton Business School
University of Southampton
Southampton, UK

Jaywant Singh
Kingston Business School
Kingston University
Kingston Upon Thames, UK

The field of luxury studies increasingly encompasses a variety of perspectives not just limited to marketing and brand management. In recent times, a host of novel and topical issues on luxury such as sustainability, counterfeiting, emulation and consumption trends have gained prominence which draw on the fields of entrepreneurship, sociology, psychology and operations.

Examining international trends from China, Asia, Europe, North America and the MENA region, *Palgrave Advances in Luxury* is the first series dedicated to this complex issue. Including multiple perspectives whilst being very much grounded in business, its aim is to offer an integrated picture of the management environment in which luxury operates. It explores the newer debates relating to luxury consumption such as the signals used in expressing luxury, the socially divisive nature of luxury and the socio-economic segmentation that it brings. Filling a significant gap in our knowledge of this field, the series will help readers comprehend the significant management challenges unique to this construct.

All submissions are single blind peer reviewed. For more information on our peer review process please visit our website: https://www.palgrave.com/gp/book-authors/your-career/early-career-researcher-hub/peer-review-process.

This series is indexed in Scopus

Alba D'Aniello •
Ana Carmela Martinez Levy •
Cesare Amatulli • Matteo De Angelis •
Rumen Pozharliev

Aesthetics and Sustainable Luxury Consumption

Evidence from Scientific Literature, Case Studies and Consumer Neuroscience

palgrave
macmillan

Alba D'Aniello
University of Bari "Aldo Moro"
Taranto, Italy

Ana Carmela Martinez Levy
Luiss Guido Carli University
Rome, Italy

Cesare Amatulli
University of Bari "Aldo Moro"
Taranto, Italy

Matteo De Angelis
Luiss Guido Carli University
Rome, Italy

Rumen Pozharliev
Luiss Guido Carli University
Rome, Italy

ISSN 2662-1061 ISSN 2662-107X (electronic)
Palgrave Advances in Luxury
ISBN 978-3-031-96832-7 ISBN 978-3-031-96833-4 (eBook)
https://doi.org/10.1007/978-3-031-96833-4

This Palgrave Macmillan imprint is published by the registered company Springer Nature Switzerland AG.
The registered company address is: Gewerbestrasse 11, 6330 Cham, Switzerland

Foreword

Luxury and sustainability have long been perceived as opposing forces in the world of consumer behavior. On the one hand, luxury is often associated with exclusivity, indulgence, and an aura of excess. On the other, sustainability advocates for resource conservation, ethical production, and a reduction of environmental and social harm. At first glance, the two concepts appear incompatible. However, this new book, "Aesthetics and Sustainable Luxury Consumption: Evidence from Scientific Literature, Consumer Neuroscience, and Case Studies," challenges this traditional dichotomy. It presents several compelling arguments that luxury and sustainability not only can coexist but also can complement each other, leading to a new conceptual paradigm. Authored by a team of scholars with extensive research experience in sustainable luxury, this book offers a rich and evidence-based exploration of the intersection between luxury goods and sustainability.

One of the central arguments of the book is that the evolving nature of luxury consumption is enabling an overlap between luxury and sustainability. Historically, luxury consumption was primarily driven by the desire for conspicuous consumption—the need to display wealth and status through extravagant purchases. However, this dynamic is progressively shifting. Today, a growing segment of consumers, particularly younger generations, is moving toward what scholars have termed "conscious consumption", including a deep interest in high-end goods with a sustainability angle. This gradual shift reflects a deeper concern for the social and environmental impact of one's choices, which is reshaping the luxury market.

Millennials and Generation Z consumers are driving this change. These younger generations demand greater transparency, traceability, and ethical production practices from luxury brands. For them, luxury is no longer defined solely by opulence but by its alignment with sustainability principles. Corporate social responsibility initiatives, once seen as peripheral, now play a central role in shaping brand perception and consumer loyalty. This generational change underscores the growing expectation that luxury products must not only be beautiful and exclusive but also ethically produced and environmentally responsible.

The book also delves into how the very meaning of luxury is being redefined in certain product categories. Innovations like lab-grown diamonds and mushroom leather illustrate how sustainability and luxury can converge. Lab-grown diamonds, for instance, offer the same brilliance and allure as mined diamonds without the environmental and ethical concerns associated with traditional diamond mining. Similarly, mushroom leather represents a sustainable alternative to animal leather, combining high-end aesthetics with eco-conscious production. These examples highlight how the luxury industry can innovate to meet consumer expectations while minimizing its environmental footprint.

One of the most striking aspects of the book is its focus on the role of aesthetics in sustainable luxury consumption. Luxury goods and service environments are inherently linked to the concept of art, and their looks and feel are key drivers of consumer desire. However, the authors argue that aesthetic choices made by luxury brands can either enhance or diminish perceptions of sustainability. Recent advancements in Neuroaesthetics—the study of how the brain perceives and processes art stimuli—offer insights into how consumers unconsciously interpret the sustainability of luxury products. By integrating findings from the literature, the book provides a novel perspective on how aesthetic dimensions can influence sustainable consumption.

I believe one of the book's most significant contributions is its exploration of how luxury brands have embraced various facets of sustainability, such as diversity, product durability, and the preservation of local traditions. For example, a compelling case study discussed in the book is Rihanna's Fenty Beauty brand, part of the LVMH Group. Fenty Beauty has redefined inclusivity in the luxury cosmetics market by offering products that cater to all skin types and tones. This approach to inclusivity extends to sustainability, as the brand emphasizes ethical production, eco-friendly packaging, and vegan ingredients. The book also examines the philosophy of Brunello Cucinelli, a luxury brand that embodies the principles of humanistic capitalism and human sustainability. Brunello Cucinelli's approach to product design combines the highest quality raw materials, specialized craftsmanship, and timeless elegance. This philosophy of "longevity by design" ensures that the brand's garments transcend fleeting trends and seasons, promoting a more sustainable approach to fashion. By prioritizing durability and timelessness, Brunello Cucinelli demonstrates that luxury can be both aesthetically pleasing and environmentally responsible. Finally, the authors also emphasize the importance of "Made-In" aesthetics in shaping consumer perceptions. Certain aesthetic dimensions, such as the use of local materials and traditional craftsmanship, are more effective in driving sustainable choices. For example, the book showcases the philosophy of Borgo Egnazia, a luxury hospitality brand rooted in the local land and culture of the Apulia region in Italy. Borgo Egnazia's approach to design and hospitality embodies sustainability through its use of local materials, such as tuff stone, which is both authentic and energy efficient. This case study illustrates how aesthetic choices, whether in products or services, can reflect a brand's commitment to sustainability while enhancing its appeal to conscious consumers.

The inclusion of these real-world cases and different interpretations of sustainability enriches the book and makes its arguments highly relevant to both academic and managerial audiences. By showcasing how leading luxury brands have embraced sustainability, the authors provide actionable insights for other brands aiming to navigate the sustainable transition. These examples also serve to inspire consumers, encouraging them to view luxury as a vehicle for positive change rather than as a symbol of excess.

In conclusion, "Aesthetics and Sustainable Luxury Consumption: Evidence from Scientific Literature, Case Studies, and Consumer Neuroscience" is a timely and essential contribution to the field of consumer behavior. It challenges the traditional notion that luxury and sustainability are incompatible and offers a new framework for understanding how the two can coexist. Through its analysis of the literature, case studies, and consumer neuroscience, the book provides valuable insights into the motivations behind sustainable luxury consumption and the aesthetic dimensions that influence it. At a time when the fast fashion industry continues to thrive, despite growing concerns about sustainability, this book underscores the importance of promoting high-end, sustainable alternatives. By cementing the idea that luxury and sustainability can go hand in hand, the book has the potential to shift consumer behavior toward more responsible and mindful consumption and is an indispensable resource for scholars, practitioners, and anyone interested in the intersection of aesthetics, sustainability, and luxury.

Columbia Business School Silvia Bellezza
New York, NY, USA
January 31, 2025

Acknowledgements

The authors acknowledge financial support under the National Recovery and Resilience Plan (NRRP), Mission 4, Component 2, Investment 1.1, Call for tender No. 104 published on 02.02.2022 by the Italian Ministry of University and Research (MUR), funded by the European Union—NextGenerationEU—Project Title "Made in Italy and sustainable development: A multidisciplinary project for showing the determining role of luxury and aesthetics" (Project No. 2022YJ9SWX)—CUP H53D23002630006—Grant Assignment Decree No. 967 adopted on 30.06.2023 by the Italian Ministry of University and Research (MUR).

Contents

About the Authors

Alba D'Aniello is research fellow at the University of Bari, in Italy currently working on a research project about Made in Italy and sustainable development focused on the role of luxury and aesthetics in shaping sustainable choices. She teaches Marketing and Qualitative and Quantitative Research Methodology for Marketing at Luiss Guido Carli University, where she is also the executive director of the X.ITE (X = Behavioral Insights Technology Enhanced) Research Center, on technologies and market behavior. The latter works as a knowledge-transfer center for professionals, firms, and institutions to manage the transition towards a tech-empowered and sustainable future. She got her PhD in Management in 2023, with a thesis focused on consumers' evaluation of sustainable food packaging. Indeed, her main research interest is transformative consumer behavior and sustainability in all its nuances, ranging from consumers' evaluations of green products, how specific product features or production processes may hinder or enhance quality inferences, as well as innovative strategies to boost more responsible choices.

Ana Carmela Martinez Levy is research fellow at LUISS Guido Carli University in Italy, working on a research project about Made in Italy and sustainable development. More particularly, with a focus on the role of luxury and aesthetics in shaping sustainable choices, with an emphasis on neuroscience research. Additionally, she teaches International Marketing at a center affiliated with the University of Granada and since 2016 she has been collaborating with BrainSigns Company, a highly reputable company specializing in applied neuroscience across multiple fields. Her expertise focuses on the application of neuroscience to consumer behavior in marketing. She attended a Master in Neuromarketing, carrying out her thesis on Neuroaesthetics; Thesis title: "Art and Neuroscience: Neuroaesthetics" in 2015. In 2021 she completed her PhD in Communication, Social Research and Marketing ("Doctor Europeaus") from the Department of Communication and Social Research, Sapienza, University of Rome, with a thesis entitled: "New frontiers in assessing consumer behavior through neuroscientific approach: a case study on alcoholic vs non-alcoholic beers". Much of her research has focused on neuroscience consumer behavior on which she has published several articles and chapter books.

Cesare Amatulli is Full Professor of Marketing at the University of Bari in Italy, where he also got his PhD. He also teaches "Brand Management", "Research Methodology for Marketing", and "Strategic Marketing in Luxury Industries" at LUISS University in Rome. He has been Visiting Researcher at the Ross School of Business (University of Michigan, USA) and at the University of Hertfordshire (UK). His research focuses on consumer behavior and more specifically on luxury consumption and sustainable luxury. His studies are mainly based on quantitative methods, specifically experiments, to underline cause-effect relationships between communication stimuli and consumers' behaviors. He has published in various international scientific journals in the fields of marketing, consumer behavior, management and psychology, such as Journal of Consumer Research, International Journal of Research in Marketing, Journal of Business Ethics, Journal of Business Research, Journal of Retailing and Psychology & Marketing. He is the author of the book titled "Sustainable Luxury Brands" published with Palgrave Macmillan.

Matteo De Angelis is Full Professor of Marketing at Luiss University, in Rome. He got the PhD in Management at the University of Bologna in 2008. At LUISS Guido Carli he is currently Coordinator of the PhD in Management and Director of the Academic Research Unit at the Luiss

X.ITE Research Center, as well as Director of the Master in Marketing Management at Luiss Business School. His work has been published in prestigious international journals in the fields of marketing, consumer behavior, management and psychology, such as *Journal of Marketing*, *Journal of Marketing Research*, *Journal of Consumer Research*, *Strategic Management Journal*, *International Journal of Research in Marketing*, *Journal of the Academy of Marketing Science*, *Psychological Science*, and many others. A good part of his research has focused on luxury marketing, particularly on sustainable luxury, on which he has published several articles, and a book titled "Sustainable Luxury Brands" with Palgrave Macmillan.

Rumen Pozharliev is Assistant Professor of Marketing at Luiss University, in Rome. He got the PhD in Marketing at the Erasmus University Rotterdam in 2017. At LUISS Guido Carli he is currently Teaching Area Leader and Senior Researcher at the Luiss X.ITE Research Center, as well as Director of the Master in Retail, E-commerce and Distribution at Luiss Business School. Currently, he is also teaching Marketing Management at Temple University. His work has been published in prestigious international journals in the fields of marketing, neuroscience, and consumer behavior such as *Journal of Marketing*, *Journal of Marketing Research*, *Journal of Consumer Research*, *Journal of Retailing*, *Journal of Advertising*, *Frontiers in Human Neuroscience*, and many others. A good part of his research has focused on the social aspects of luxury marketing and consumer neuroscience on which he has published several articles, and a book titled "La Mente del Consumatore" with Patrizia Cherubino.

List of Figures

List of Tables

1

Introduction

The luxury business is increasingly playing a crucial role for sustainable development. Traditionally linked to exclusivity, magnificence, excess, and indulgence, luxury is currently experiencing a paradigm shift that incorporates values of minimalism, inclusivity, and sustainability. At the same time, the evolving values of modern consumers, especially among younger generations, are undermining the conventional foundations of luxury. Consumers are increasingly demanding openness, authenticity, and purpose. Although apparently oxymoronic, a growing number of scholars agree on the idea that luxury is sustainable by definition: luxury products and experiences have historically been distinguished for their craftmanship, high-quality, sophistication, and beauty, which in turn affect their functional endurance over time (Amatulli et al., 2017).

The present book proposes a novel perspective on sustainable luxury consumption emphasizing the key role of aesthetic dimensions in shaping responsible choices. Indeed, luxury goods and service environments are beautiful, refined, and aesthetically pleasing, which makes them intrinsically linked to the concept of "art" (Kapferer, 2014). However, both luxury and aesthetics are subjective constructs and different aesthetics choices made by luxury brands can either strengthen or decrease

A. D'Aniello et al., *Aesthetics and Sustainable Luxury Consumption*, Palgrave Advances in Luxury, https://doi.org/10.1007/978-3-031-96833-4_1

consumers' perceptions of product sustainability, both consciously and unconsciously. Through the analysis of relevant literature, as well as drawing on recent unpublished empirical research, we present a consumer-based view of sustainable luxury consumption according to which aesthetics is a fundamental driver of the luxury customer experience, which in turn may affect product durability, above and beyond its functional quality. Indeed, when aesthetic attributes coincide with the tenets of environmental stewardship and social equality, they construct a compelling narrative: one in which enjoyment and responsibility are not merely compatible but also mutually reinforcing.

This book offers an extensive examination of these subjects, integrating theoretical perspectives with case studies and managerial insights. Accordingly, it is structured into five chapters. In Chap. 2 we define luxury and analyze the evolving landscape of luxury consumption, emphasizing how changes in consumer motivations and emerging trends such as mindful consumerism are influencing the demand for luxury products and services. Then, in Chap. 3 we explore the apparent tensions between sustainability and luxury. We argue that quality, heritage, and craftmanship are key to reconcile the two concepts and it presents the "Made in" as a fundamental aspect of authenticity and sustainability. Starting from Chap. 4, we introduce the field of aesthetics, focusing on its relevance for the luxury sector, and arguing that beauty affects emotions, decision-making, and value generation. Through the introduction of Neuroaesthetics, we elaborate on how aesthetic design may help sustainability by enhancing enduring consumer connection and product valuation. Furthermore, in Chap. 5 we explore the convergence of aesthetics and sustainability within the Italian luxury sector, focusing on both products and services. Through the analysis of critical sectors—i.e., fashion, jewelry, watches, automobile, hospitality, and fine dining—we illustrate that Italian firms incorporate sustainability and purpose, while maintaining heritage and excellence, also through aesthetic choices. Finally, in Chap. 6 we bridge together the book's ideas to offer a prospective outlook, proposing a novel theoretical framework that integrates sustainability, aesthetics, and consumer behavior, detailing the relationships among these constructs. Highlighting managerial insights and future research directions, we hope that the present work may stimulate future scholars and practitioners to further develop this fascinating field, given its theoretical and practical relevance for a real sustainable production and consumption. The convergence of aesthetics and sustainability is not simply a strategic benefit for companies but a moral obligation for an industry capable of shaping and influencing global cultural standards. This transformation represents both a challenge and an opportunity for luxury brands: to innovate while preserving their past, to modernize aesthetics without undermining ethics, and to inspire without exhausting resources.

References

Amatulli, C., De Angelis, M., Costabile, M., & Guido, G. (2017). *Sustainable luxury brands: Evidence from research and implications for managers*. Springer.

Kapferer, J. N. (2014). The artification of luxury: From artisans to artists. *Business horizons, 57*(3), 371–380.

2

Luxury Marketing: State of the Art and Evolving Consumer Behaviors

The luxury industry represents a unique opportunity for firms capable of offering excellent products and services. During significant geopolitical and macroeconomic turbulence, the luxury industry proves to be an extremely profitable and relevant sector; according to the annual Bain & Company-Altagamma Luxury Study, in 2023 the global luxury market reached €1.5 trillion, establishing a new record of growth (D'Arpizio et al., 2024). Indeed, after a post-pandemic decline which also affected luxury expenditures (including physical goods, services, and experiences), global wealth is rising again: billionaires have recovered from net losses of 2022 (Foster, 2024) and the global luxury industry even exceeded its pre-pandemic level (D'Arpizio et al., 2024).

What has been defined as a "*second Gilded Age*" (Foster, 2024) makes the luxury sector particularly attractive for companies that can (1) create, communicate, and deliver superior-quality products or services (Amatulli et al., 2017) and (2) effectively manage brand equity over time (Keller, 2009): in 2022, the 100 largest luxury goods companies generated sales of $347 billion (+42 billion compared to 2021), and the top 100 players saw overall growth of 20% (Faccioli et al., 2023).

A. D'Aniello et al., *Aesthetics and Sustainable Luxury Consumption*, Palgrave Advances in Luxury, https://doi.org/10.1007/978-3-031-96833-4_2

However, the luxury market is not just going through an era of expansion, but it is rather evolving: the overall structure of the sector, as well as the tactics that marketers must use to attract and retain their target audience are strongly influenced by the ever-evolving demands of the customers. Because of this, understanding the luxury industry necessitates a thorough examination of the various trends that might change consumers' perceptions and actions in relation to luxury consumption choices. The present chapter seeks to provide an overview of the main characteristics of luxury brands, describing the unique features of products and services, subsequent marketing strategies, as well as the emerging tastes of luxury consumers segments, which in turn would help to identify and elaborate on the main challenges that luxury marketers are facing nowadays.

Definition and Unique Characteristics of Luxury

Due to its sensitivity to different cultural and temporal shifts, the concept of luxury is particularly diversified, and complex. Indeed, the marketing literature lacks a singular definition of luxury and for three main reasons. First, luxury is a *relative and subjective* concept that reflects individual and societal beliefs. In other words, luxury might mean different things to different people or even to the same person (Berthon et al., 2009): an individual's perception of goods and services might change proportionally to an increase in their own standard of living (Amatulli et al., 2017). Second, as the aspirations and goals of both current and prospective customers change over time, so does the way in which luxury is conceptualized. Third, the concept of luxury is cross-sectorial, as it relates to different product and service categories.

Nonetheless, one of the simplest definitions of luxury has been provided by Adam Smith (1776) who classified consumption into four categories: *necessities* (needed to sustain life), *basics* (for normal growth and prosperity), *affluence* (items not required for growth and prosperity), and *luxury* (items that are rare or difficult to obtain, and expensive). Both rarity and high price have been widely accepted by scholars as key characteristics of luxury, which have been further enriched over time with additional features that we briefly summarize below (e.g., Amatulli et al., 2017; Dubois et al., 2001, 2005):

1. *Excellent quality*. Quality in luxury refers to both input materials and production processes, which draws on artisan traditions. Making a luxury product starts with carefully selecting prestigious raw materials that are then assembled by highly skilled employees. Indeed, luxury products are often the outcomes of artisan craftsmanship and generational traditions, which together allow major brands to maintain high-quality standards over time. Moreover, those materials and traditions are commonly associated with specific regions, which leads to a country-of-origin and "Made In" effect that enhances luxury products' desirability. For example, Tuscany is well-known for its leather district in Italy for high-end leather jackets and accessories. Similarly, the high-end Italian fashion brand Brunello Cucinelli is recognized as an "*artisanal industrialist*", since its business model draws on highly specialized craftsmen factories, mainly located in Umbria: almost 52% of the collections result from the manual component (see also Chap. 4 – "The gentle luxury of Brunello Cucinelli").[1]
2. *Scarcity and exclusivity*. Consumers of luxury products and services expect them to be *rare* or even *unique* and this is a main driver of their attractiveness for traditional luxury consumers (e.g., Rosendo-Rios & Shukla, 2023). While scarcity is a natural consequence of utilizing prestigious raw materials (which are limited in nature) and excellent artisanship, it can also be artificially caused by sophisticated use of marketing tactics. Indeed, a distinctive trait of luxury businesses is their propensity to intentionally create a sense of *exclusivity*: for example, through limited production and selective distribution channels. Both strategies convince target consumers that the item is one-of-a-kind, thereby justifying high prices and years-long waiting lists. This is exactly the case for most iconic products such as the forty-year-old Hermès Birkin bag or a pair of Chanel Slingbacks. The most success-

[1] Brunello Cucinelli Business Model – Italian Craftmanship and manual skills. Available at https://investor.brunellocucinelli.com/en/business-model/italian-craftmanship-and-manual-skills.

ful luxury firms have found the right balance between exclusivity for actual clients and accessibility for aspirational consumers.

3. *Aesthetic appeal*. Luxury brand managers work hard to create pleasant and aesthetically sophisticated experiences, regardless of whether they are selling physical products or intangible services. Indeed, luxury offers are usually extremely attractive to customers, and sometimes considered "works of art" due to their beauty and refinement (Venkatesh et al., 2010). Prior studies identified aesthetics as a significant attribute for luxury goods and services (e.g., Holmqvist et al., 2020) as well as the central role of the recipients and their aesthetic taste and refinement (Berthon et al., 2009). According to Dion and Arnould (2011), aesthetic appeal is strictly linked to the concept of high quality, since luxury should simultaneously embrace "*aesthetic and technical excellence*" (p. 504). For that reason, the "Made in Italy" industry, often described as being *beautiful and well-made*, represents a unique domain for a deeper analysis of the aesthetic dimensions which characterize a luxury product. Luxury businesses should focus on establishing a distinctive aesthetics that is attractive to their target customers and conveys a sense of exclusivity and uniqueness.
4. *Heritage and history*. Major luxury brands can claim centuries of history: Just consider Hermès (French) founded in 1837, Cartier (French) founded in 1847, Omega (Swiss) and Burberry (British) founded in 1848, Louis Vuitton (French) founded in 1857, Bulgari (Italian) founded in 1884, Rolex (Swiss) founded in 1905, Gucci (Italian) founded in 1921, Ferrari (Italian) founded in 1947, and Brunello Cucinelli (Italian) founded in 1978. These brands' long histories heighten the authenticity and prestige of their offerings. That heritage is also commonly linked to the brand's country-of-origin: For example, the long history of worldwide success for high-end brands such as BMW (1916) and Porsche (1931) have imbued Germany with a heritage of excellent quality, precision, and efficiency. Similarly, France and Italy are recognized worldwide for their long history of success in the fashion industry: the Deloitte "*Global Powers of Luxury Goods 2023*" report identifies Italy as the leading country of luxury worldwide—with 23 firms in the top 100 ranking—and France—with 4 firms in the top 10—the country with most solid financial perfor-

mances, especially due the fact that it hosts big luxury groups such as LVMH and Kering (Faccioli et al., 2023).

5. *Fulfilling a dream.* Luxury consumption is an inherently symbolic act that fulfills social needs (Dubois, 2020). Vickers and Renand (2003) suggest that luxury goods' primary value is psychological, offering a distinctive mix of functionalism, experientialism, and symbolic interactionism compared to non-luxury goods. Drawing on Karl Popper's "three worlds", Berthon et al. (2009) defined three dimensions of luxury brands as follows: (1) *functional (objective-material)* reflects the product's objective attributes and how it satisfies specific needs; (2) *experiential (subjective-individual)* captures the subjective evaluation of consumers' individual taste and the hedonic value they attach to luxury consumption; (3) *symbolic (collective-social)* deals with the social nature of consumption that is reflected in the meaning and values attached by others. In other words, luxury consumption allows consumers to dream by communicating their own identity, as well as their *actual* or *desired* status (Dubois, 2020). Moreover, the luxury product itself may become an object of dreams, and this is especially true for the most iconic items associated with the brand: consumers may first daydream about the moment in which they will be wealthy enough to indulge buying a timeless masterpiece (Kapferer, 2017) such as the Bulgari B.zero1 ring,[2] or a red sartorial dress by Valentino Garavani,[3] that will never go out of fashion. As a result of their continued production, these items incorporate story, art, and tradition, which stimulate desires of the aspirational consumer segments (Kapferer, 2017).
6. *Very high price.* All the above-mentioned features allow luxury companies to charge significantly higher prices for their products relative to the mass market; those prices then help nourish the associations with aesthetics, quality, heritage, exclusivity, and dreaming. When setting the appropriate pricing strategy, luxury brand managers should not only consider production costs, but also the intangible values associ-

[2] Bulgari icons: https://www.bulgari.com/it-it/bvlgari-icons.html.

[3] Valentino Icons Only, Rosso Valentino: https://www.valentino.com/it-it/experience/icons-only-rosso-valentino.

ated with the brand and transferred to the product itself. Analogously, "*pure*" luxury brands never offer promotions or special offers (Kapferer, 2017) and luxury consumers are even insensitive to price increases: According to the 2023 Bain & Company–Altagamma Luxury Market Study, almost all personal luxury goods categories observed a price increase, which only partially undermined sales volumes, thus leading to significant economic growth (D'Arpizio et al., 2024).

Similarly, Ko et al. (2019, p. 406) proposed a consumer-based definition of a luxury product, drawn from the analysis of the extant literature, according to which an ideal luxury product should always display the following characteristics: "*(1) being of high quality; (2) offering authentic value via desired benefits, either functional or emotional; (3) having a prestigious image within the market built on qualities such as artisanship, craftsmanship; (4) being worthy of commanding a premium price; and (5) being capable of inspiring a deep connection, or resonance, with the consumer*".

Peculiarities of Luxury Services

Even though most of the above characteristics can be easily applied to luxury services and experiences, we underline that the intangibility, heterogeneity, inseparability, and perishability (Zeithaml et al., 1985) that typically characterize services require to consider additional elements of the marketing mix, namely People, Process, and Physical facility (Booms & Bitner, 1981) and distinctive customer needs.

For that reason, Wirtz et al. (2020) proposed a careful analysis of luxury services. We report here some of the most relevant peculiar characteristics of luxury services:

- Customers of luxury services cannot gain pleasure from owning high-quality products. Therefore, they express psychological ownership by recalling the experience and sharing it with others, either in face-to-face interactions or through posts and stories on social networks.
- Most of all, luxury firms should emphasize the psychological benefits accrued from enjoying the extraordinary luxury service experience. To

illustrate, luxury firms should educate their customers towards a deeper understanding of the level of sophistication of the experience they are living (e.g., Eisingerich & Bell, 2008) or prompt them to share these memories on their social networks. Moreover, services are heterogeneous (Zeithaml et al., 1985), thus luxury service managers could emphasize the fact that every single experience is unique and personalized.

- Indeed, most service experience happens in real-time, through direct interaction with the customers who can even co-create the output.
- The perception of services as luxurious can be fragile. Even if services managers try to "*make tangible the intangible*" (i.e., enhance the salience of the luxury experience) by tightly managing various elements (e.g., the physical facility, employee training, standardized processes, but also pictures, stories and souvenirs, etc.), customers' entire perception of the experience can be undermined by even a minor negative aspect.
- Finally, services can be designed with various degrees of exclusivity and luxuriousness, that heavily depend upon consumers' subjective evaluations: A luxurious resort in an exotic destination or a Michelin-starred dinner could appear as "ordinary" to the wealthiest consumers.

To sum up, luxury services can be defined as "*extraordinary hedonic experiences that are exclusive*"—i.e., difficult to afford (monetary exclusiveness), difficult to access (social exclusiveness), and even difficult to enjoy (hedonic exclusiveness)—whose evaluation strictly depends on subjective consumers evaluation as well as objective services attributes (Wirtz et al., 2020, p. 682).

Managing Luxury Brands

The American Marketing Association (AMA, 2017) defines marketing as "*the activity, set of institutions, and processes for creating, communicating, delivering, and exchanging offerings that have value for customers, clients, partners, and society at large*". Value creation requires a careful understanding of customer needs and adopting coherent marketing strategies.

Establishing long-term profitable relationships with clients is particularly relevant for luxury firms. Nevertheless, these firms face unique challenges: the fundamental principle of traditional marketing is to address client needs with explicit and appropriate answers, while the fundamental principle of luxury marketing is to initiate trends, create new demands, and even identify and offer new luxuries (Amatulli et al., 2017). Indeed, in the context of luxury brand management, the traditional approach to economic equilibrium is reversed, since luxury firms aim at generating an abundance of demand without fulfilling it: this strategy allows brand managers to increase price and generate higher profits by intentionally reducing the supply in the market (Kapferer, 2017). These additional profits can then be used to invest in building a strong and respected brand image.

Consumers form their opinions of luxury based on both the tangible and intangible characteristics of luxury products and services. According to Keller (2009), luxury brands can be considered the purest form of branding. In today's particularly competitive setting, companies have significant influence in shaping and spreading their image. They can then use this influence to establish trends, affect consumer demand, and build brand worth. Nevertheless, companies require advanced managerial expertise to stimulate demand for superfluous and expensive products. Regarding this matter, Keller (2009) presented a valuable classification of ten characteristics of luxury brands, which are summarized in Table 2.1.

Luxury companies need to carefully manage all these interconnected features to fully distinguish themselves from mass-market players (Amatulli et al., 2017). In other words, premium brand management is extremely specialized and challenges traditional marketing principles—what we can define as the "anti-laws" of luxury marketing, which can be translated in some practical rules to nourish growth while keeping luxury brands' exclusive image (e.g., Kapferer & Bastien, 2009; Kapferer, 2016; Kapferer, 2017):

1. Selling to "mass-niches": scale up the production of small, limited-edition collections to create a sense of exclusivity and make purchasing a priority for clients.

Table 2.1 Keller's taxonomy of luxury brand features

Luxury brand features	Description
Premium image, strictly controlled	A luxury brand's success depends on its premium image, which should be reinforced by company prestige and product features. Managers should only promote values, associations, and brand extensions that are coherent with the established brand image.
Intangible associations and aspirational image	Successful brands can form positive associations by leveraging histories, places, and heritage. The aspirational image attracts prospective customers who dream about luxurious consumption and status.
Marketing tactics for high-quality and pleasurable experiences	Luxury brands manage every aspect of their interaction with consumers to convey a superior-quality image and ensure customers' satisfaction.
Brand elements as drivers of brand equity	Luxury brands need to carefully manage their various elements—logos, colors, symbols, packaging, signage, etc.—to differentiate themselves from other market players.
Exclusive secondary brand associations	Luxury brands can reinforce their image through secondary brand associations with celebrities, exclusive events, and country of origin effects. For instance, if luxury managers want to convey elegance and beauty as brand qualities, then they should carefully select and coordinate with celebrities who can express those values.
Selective distribution	Luxury managers should prefer direct distribution channels and flagship stores so that they can control all the relevant aspects of merchandising, pricing policies, sales force behavior, and interior design.
Premium pricing and few discounts for high quality	Luxury brands should pursue production excellence and communicate the higher value of their products to their target customers.
Careful management of brand architecture	To balance exclusivity with accessibility, luxury brands often promote lower-priced offerings to attract new prospective customers.
Broadly defined competition	Luxury brands compete with not only other luxury brands (even if they belong to different product or service categories), but also mass-market brands in their industry.
Legal protection of trademarks	Luxury products are vulnerable to counterfeiting. Brand managers should protect their brand(s) in as many ways as possible.
Aesthetic value	Luxury products and services are carefully designed to be pleasant and aesthetically appealing.

Source: Authors elaboration from Amatulli et al. (2017); Keller (2009)

2. Further elevating luxury to a "work of art" through secondary associations and partnerships: collaborate with renowned, innovative artists to enhance the limited collection's worth and stimulate the potential secondary market for that item. Also, firms could periodically offer amazing and distinctive creations made by renowned designers, available for purchase at remarkable costs, through charity auctions that generate excitement.
3. Limit distribution: luxury goods need to be sold in a few, preferably company-owned, stores, since the perceived prestige is inversely correlated with the number of stores. This can also be considered as a way to sustain the "luxury dream" by constantly keeping the penetration rate below the brand awareness rate.
4. Exploit pricing strategies: luxury brands should be expensive enough to be considered a worthwhile sacrifice for their buyers. Firms should promote an ever-increasing pricing strategy, which may act as a counterbalance to the progressive spread of wealth and luxury items in the mass market.

Luxury Consumers: Underlying Motivations and Consumption Approaches

Scholars have deeply explored consumers' motivation for luxury consumption, which are far away from the rational choice theory (Scott, 2000). Indeed, from the classic economists' standpoint, consumers usually make choices based on a thorough assessment of all options' costs and benefits. Nevertheless, it is evident that rationality alone cannot fully elucidate the intricate motives behind luxury purchases (Kapferer, 2017): consumers are willing to pay a premium price and are insensitive to price increases, mainly because they look for intangible (i.e., symbolic and experiential value) rather than tangible value (Leibenstein, 1950; Kapferer, 2017). Hence, as luxury consumption usually emphasizes experiential, emotional, and symbolic values over functional ones, it is often described as superfluous. Analogously, based on the well-known human needs' hierarchy proposed by Maslow (1943), human needs can be represented as a pyramid of five tiers: the bottom level includes basic physiological needs such as feeding, hydration, and rest; the second level pertains to the need for safety and security, like health, employment, or having a safe house; then, the third tier refers to the need for belonging, encompassing the need for love, companionship, and social approval; esteem needs such as need for accomplishment and recognition from others come next, followed by the highest level, which includes the self-actualization needs, which pertain to moral development and the recognition of their maximum capabilities. Undoubtedly, luxury consumption mostly satisfies the higher-level needs, as it often enables consumers to enhance their self-esteem or project a favorable image of themselves. Furthermore, luxury objects play a significant role in helping individuals achieve their goals by providing them with a better understanding of themselves, thus also satisfying their need for self-accomplishment (Amatulli et al., 2017). Furthermore, consumers may perceive that they are fulfilling their need for belonging by making ostentatious choices. This perspective aligns with the work of Leibenstein (1950) who identifies three main external effects of consumption, which further increase the perceived utility drawn from the product/service purchased: because of the "Bandwagon effect", consumers are motivated to buy non-necessary products because of their need to conform to other people they desire to be associated with (e.g., Kastanakis & Balabanis, 2012; Vigneron & Johnson, 1999). On the opposite side, the "Snob effect" refers to consumers' need for uniqueness, which can be expressed as the desire to differentiate themselves from another reference group (e.g., Eastman et al., 2018). Finally, the "Veblen effect" is observed for conspicuous consumption, when consumers place a higher emphasis on price as a measure of social status, since they aim to impress others through their choices (e.g., Bagwell & Bernheim, 1996). Drawing on this conceptualization of potential interpersonal effects of consumption choices, Vigneron and Johnson (1999) propose two additional personal effects, namely: (a) the "hedonic effect", which is related to individual's emotions driven by the enjoyment of the consumption experience; and (b) the "perfectionism effect", according to which consumers buy luxury products as they are high-quality seekers. Finally, as the things we own have a significant role in shaping and representing our sense of self (Belk, 1988), we further propose a third personal effect, which we will refer to as "self-effect":

consumers buy luxury products/services because of their need to strengthen their own self-concept (e.g., Wiedmann et al., 2009; Gil et al., 2012), whether actual or desired (Belk, 1988). Those who possess an independent self-concept will focus on hedonic, utilitarian, and self-communication objectives (Ko et al., 2019). On the contrary, consumers with a strong interdependent self-concept will place greater importance on the interpersonal effects of luxury buying (Ko et al., 2019).

In a nutshell, luxury purchase can be attributed to both internal and external reasons (Amatulli et al., 2017). The *externalized* approach to consumption occurs when individuals seek luxury brands primarily to gain the acceptance of others and are therefore influenced by the ideas and opinions of third parties; the *internalized* approach, instead, refers to consumers' desire to experience positive emotions and align with their particular taste and style (Amatulli & Guido, 2012).

Table 2.2 summarizes all the above-mentioned perspectives bridging together consumers' motivation and needs satisfied through luxury consumption, since they are producing interpersonal or personal effect, thus being driven by an internalized or externalized approach to luxury purchases.

Common Segmentation Schemes for Luxury Buyers

The distinction of luxury consumers based on their primary motivation to purchase luxury items represents the basis of different value-based segmentation schemes that are particularly popular within the marketing literature. Indeed, while the prevailing indicator for luxury expenditure has been restrained to consumers affluence (also see the following section about contemporary luxury buyers for an example), going beyond demographic and economic indicators allows to better understand the prevailing reasons for luxury buying (Ipsos, 2022).

For example, Dubois et al. (2005) distinguish luxury buyers based on their attitudes towards such products: *Elitists* perceive luxury as a means for exclusive consumption by a limited audience who possess adequate financial means. According to them luxury is for persons who possess sufficient education, discernment, and sophistication to genuinely value it.

Table 2.2 Luxury consumption: approaches, motives, and desired effects

Approach (Amatulli & Guido, 2012; Guido et al., 2020)	Classification	Explanation	Effects type (Vigneron & Johnson, 1999)	Main references
Externalized	**Bandwagon effect/ social comparison**	Consumers are motivated to purchase a product/service because others are doing the same. It is a matter of conforming to people they want to be associated with.	**Interpersonal**	Leibenstein (1950)
	Snob effect/need for uniqueness	Consumers are motivated not to buy a specific product/service because others are doing the same. It is a matter of feeling unique, exclusive, diverse.		Leibenstein (1950) Snyder and Fromkin (1977)
	Conspicuous consumption/ Veblen effect	Consumers are motivated to buy a specific product/service because of its higher price. It is a matter of impressing others.		Leibenstein (1950) Han et al. (2010)
Internalized	**Self-effect (self-concept and the extended self)**	Consumers buy luxury products/services to strengthen their own self-concept (either actual or desired).	**Personal**	Vigneron and Johnson (1999) Belk (1988)
	Hedonic consumption/ emotional values	Consumers buy luxury products/services driven by the desire to enjoy the shopping experience (driven by emotions)		Hirschman and Holbrook (1982)
	Perfectionism effect (Quality seekers)	Consumers look for prestige products and services as they are motivated by high quality standards		Vigneron and Johnson (1999)

Source: Authors elaboration from Amatulli et al. (2017); Ko et al. (2019); Vigneron and Johnson (1999)

Democratics propose a different perspective: according to them, luxury buyers should no longer possess specific qualifications or characteristics. Therefore, the main difference between these two segments consists in the level of accessibility of luxury goods and services. Finally, the last segment is the one of *Distants*, who have a negative view of luxury and are not attracted by it.

According to Wiedmann et al. (2009), luxury consumers can be divided into four groups: "materialists", "rational functionalists", "extravagant prestige-seekers", and "introvert hedonists". *Materialist* consumers value wealth and possessions, while quality and self-identity are not vital to them in luxury consumption. *Rational functionalists* associate luxury with quality and self-identity. *Extravagant prestige-seekers* value prestige and extravagance because of the social value of their consumption choices (i.e., they prefer distinctive and recognizable things). *Introvert hedonists* prioritize hedonism, viewing luxury as a source of pleasure and progress in life.

Han et al. (2010) proposed a segmentation of luxury buyers based on brand prominence (i.e. the visibility of the logo on luxury items). Indeed, they first distinguish between "loud" and "quiet" luxury goods: the former are recognizable products with highly visible logos, while the latter are discrete products with mostly hidden logos. Then, they identify four customer segments: Patricians, Parvenus, Poseurs, and Proletarians. *Patricians* are affluent consumers that are willing to pay more for subtle branded things that only their peers would recognize. *Parvenus* are nouveau riches who tend to exhibit their moneys, thus preferring things with conspicuous visual signs. *Poseurs* would like to signal their status through luxury consumption, but they lack financial resources. Therefore, they often buy counterfeited "loud" items. Finally, *proletarians* are lower-class people who are not interested in buying luxury, as they are not status-seekers.

Recent research by Ipsos (2022) distinguishes between six segments of affluent buyers, as follows:

(1) **Fashionistas** are enthusiastic followers of fashion, constantly seeking the latest and most desirable items to purchase. Therefore, buying high-end goods is a way to prove their trendiness and they gain plea-

sure from owning new products, thus also engaging in impulse buying behavior.

(2) **Exclusives** are status-seekers, and they tend to buy luxury mainly because of its prestige and rarity. For that reason, they want to get more knowledge and be informed about innovative and unique offers, so that they can be particularly attracted by limited editions and remarkable co-branding initiatives.
(3) **Indulgents** buy luxury items or services as a form a self-reward, to commemorate significant achievements, even if they recognize they don't need it. Moreover, as they perceive luxury items to be expensive, they derive a huge sense of gratification and are mainly driven by emotional motives.
(4) **Classics** prefer to acquire enduring and timeless products, and they attribute significant importance to heritage and stories. They carefully evaluate their purchases and favor quality over quantity. Consequently, they also tend to transmit their luxury items across generations.
(5) **Essentials** view luxury purchases as investments and recognize the inherent worth that luxury products possess. However, they tend to consider it as superficial, thus they feel a little skeptical about being habitual luxury shoppers.
(6) **Immaterials** perceive luxury as superficial and without value, so they do not spend much to buy luxury items. Even though they are more concerned about saving money and investing their finances, they prefer indulging in extravagant activities such as fine dining and global travel, thus being an extremely interesting segment for luxury service providers.

Table 2.3 summarizes the most popular segmentation criteria for luxury buyers.

In general, luxury brand managers should do their best to address the unique needs of a highly diversified customer base. To conclude this chapter, in the next paragraph, we will examine several emerging market trends that contribute to an even more intricate competitive landscape.

Table 2.3 Well-known segmentation criteria for luxury consumers from academic literature

Richards and Sturman (1977)	
Fashionable	Looking for modernity and new trends, oriented to social acceptance
Conservative	Looking for self-fulfillment and gratification, derived from excellent quality and details.
Dubois et al. (2005)	
Elitist	Positive view of luxury, that is a way to signal status. Luxury can be consumed only by those who can afford it and have enough taste and education to appreciate it.
Democratics	Positive view of luxury, which can be accessed by everyone regardless of personal requirements.
Distant	Not attracted by luxury consumption which is seen as useless and expensive.
Wiedmann et al. (2009)	
Materialists	Prioritize the values of accumulating possessions. For them, luxury consumption does not place much importance on quality and self-identity, neither try to impress relevant others through their properties.
Rational functionalists	Prioritize utilitarian values of high quality, uniqueness and building self-identity.
Extravagant prestige-seekers	Primarily prioritize social characteristics while evaluating the worth of luxury, so they look for extravagant products that clearly communicate their prestige to others.
Introvert hedonists	Focus on hedonism and life enrichment. Luxury items are primarily seen as means to gain pleasure and enhance the quality of one's life.
Han et al. (2010)	
Patricians	Wealthy individuals, willing to spend extra money for things that have subtle branding. These products serve as a way for them to show their status to other patricians.
Parvenus	Have considerable money but lack the expertise needed to comprehend subtle cues. They prefer buying "loud" luxury products, since their priority is to align themselves with other individuals who possess wealth and social status, without making distinctions between elites and newly wealthy.
Poseurs	Buy luxury goods to achieve status and differentiate themselves from less affluent individuals, but they do not have enough financial resources. Consequently, they are particularly inclined to purchasing counterfeit luxury products.

(*continued*)

Table 2.3 (continued)

Proletarians	Less rich individuals. They lack motivation to consume for the sake of social status.
Amatulli and Guido (2012); Guido et al. (2020)	
Externalized	When making luxury consumption choices individuals are driven by materialism and ostentation, valuing highly visible goods to communicate their wealth to others and gain social status.
Internalized	When making luxury consumption choices individuals are driven by individual tastes and emotions, valuing quality and heritage to fulfil personal and intimate goals.

Source: Authors elaboration

Contemporary Luxury Buyers

Due to the globalization of luxury demand, the factors influencing consumer behaviors and preferences have multiplied in complexity and variety, as even closely related countries like Italy, Spain, and France exhibit significant differences in consumption drivers (e.g., Hennigs et al., 2012). Furthermore, in recent years, luxury is getting considerably more prominent as technologies allow consumers to gain more information about the desired luxury brands, as well as the ways in which consumers might access luxury products and brands (e.g., Palombella & Faccioli, 2023).

The Boston Consulting Group and Altagamma analyzed consumption behaviors in the top 12 luxury markets worldwide—namely Brazil, China, France, Germany, India, Italy, Japan, the Kingdom of Saudi Arabia (KSA), South Korea, United Arab Emirates (UAE) , United Kingdom (UK), and United States (USA)—to provide an up-to-date snapshot of the sector and to identify emerging market needs and opportunities (BCG-Altagamma, 2023). The report highlights that China and Middle East (KSA and UAE) present promising business opportunities for the luxury sector in the post-pandemic environment: China is experiencing a huge growth in local demand for luxury, both because of travel restrictions and the progressive drop of advantages of buying abroad; similarly, while domestic development in the Middle East is expected to pick up speed, this is also anticipated to whet the spending appetite of a new generation of luxury consumers (BCG-Altagamma, 2023).

Table 2.4 Clusters of luxury customers

Cluster	Size billion €	Market weight %
Beyond money (50k€)	38	4.2%
Top absolute (20k€)	53	5.8%
Absolute & entry absolute (5k€)	260	28.8%
TRUE-LUXURY	**352**	**39.11%**
Top aspirational (2k€)	71	7.8%
Other aspirational (<2k€)	476	52.8%
TOTAL LUXURY CONSUMERS	**900**	**100%**

Source: Authors elaboration from BCG-Altagamma (2023)

In general, even though the two wealthiest consumers' clusters, which they classify as "Beyond money" and "Top absolute", represent less than 1% of the global luxury consumers, they account for almost 10% of global sales. On the opposite side, the 88.4% of "other aspirational customers" only account for 52.8% of sales. The distribution of luxury expenditures across different clusters is displayed in Table 2.4 (BCG-Altagamma, 2023).

Furthermore, nowadays the market is characterized by multigenerational complexity: Generation Y and Z are entering their top earnings years and by 2030 they are expected to account for almost the 50% and the 25% of the market respectively (D'Arpizio et al., 2024). These younger consumers demand seamless shopping experiences, which are personalized and authentic and ideally blend the virtual and physical worlds (The Economist, 2023). Moreover, especially Gen Z is well-known for being at the vanguard of societal and cultural shifts and inspiring the value systems of future generations (Atkinson & Kang, 2022; D'Arpizio et al., 2024). While older generations (mainly Gen X and Boomers) are more interested in service experiences (i.e., hotels, fine dining, wellness, etc.), Gen Y and Z attach more value to sustainability, new ownership models put forth by the sharing economy, as well as in-store tech-empowered experiences: firms should be able to navigate the transition by addressing the unique needs of their diversified customer base through specific strategic approaches for each market segment (BCG-Altagamma, 2023; D'Arpizio & Levato, 2024; D'Arpizio et al., 2024).

Emerging Trends in Consumer Behaviors

As previously said, the luxury industry is highly affected by shifts in society, time, and tradition. Originally, luxury was the outward manifestation of inherited social stratification, where the royals, clergy, and aristocrats had a social duty to display their wealth, engaging in extravagant and conspicuous expenditures. As the rigorous boundaries of social classes disappeared over time, luxury still acts as a way of reproducing social disparities, creating distance and signaling status (Bellezza, 2023; Eckhardt & Bardhi, 2020). Indeed, according to Kapferer and Bastien anything that can signal social status can be considered a luxury, even if wealthier consumers may consider it as ordinary (p. 70). From a sociological perspective, individuals tend to accept these "artificial" inequalities when they are perceived as well-deserved and self-made (Amatulli et al., 2017). All the factors discussed in this chapter, including the shifting geography, demographics, and consumption habits of luxury customers, indicate that luxury customers are not passively waiting for high-end brands to define their needs as before (Amatulli et al., 2017): especially in most mature markets, such as the Western economies, consumers are engaging with luxury products with greater restraint, careful evaluation, and discernment.

Indeed, globalization, market segments, digital communications, international travel, and culture convergence have led to a series of business challenges that luxury practitioners have never seen. Even if the world of luxury is still strictly associated with the concept of "dream", in reality the market observed a continuous process of democratization of luxury (Silverstein & Fiske, 2003; Kapferer, 2016; Bellezza, 2023) driven by rising incomes and developing economies (BCG-Altagamma, 2023; D'Arpizio et al., 2024; Foster, 2024), liquid consumption (Bardhi & Eckhardt, 2017), and the information economy (Amatulli et al., 2017) which led to the emergence of both new product lines (e.g., the so-called masstige, Truong et al., 2009; Kumar et al., 2020) and new business models such as renting (Gong et al., 2022; Pantano & Stylos, 2020; Vogel et al., 2019) and second-hand (Ki et al., 2024; Tangri and Yu, 2023) that need to be managed carefully. For example, Internet and the social media provide unrestricted access to brands, peer-to-peer communication, product and service evaluations, as well as new tools to exert influence (e.g. influencer marketing). At the same time, high technology allows for new ways to engage customers: for example, generative AI could be exploited to personalize and enhance customer experiences in digital and physical stores. Brands should be prepared to take advantage of forthcoming opportunities, including gamified, immersive, and community-based experiences that can enhance consumer engagement (BCG-Altagamma, 2023).

Luxury brand managers must effectively address emerging trends that have the potential to significantly disrupt the sector (Kapferer, 2017). Brand adaptability will be critical in capitalizing on the growth potential of new luxury consumers in general.

Conclusion

To summarize, contemporary luxury managers face the challenge of balancing the increasing worldwide access to luxury products with the imperative to maintain the distinctiveness and status that are generally associated with luxury goods. To put it differently: luxury companies face the potential danger of weakening their brands and damaging brand value if they excessively promote them in the market. However, they also risk losing their position in the global luxury market if they fail to adapt to the growing trend of luxury becoming more accessible to a wider audience.

New business models, emerging in response to societal shifts, technological developments, and environmental harms, are challenging the traditional tenets of luxury (e.g., Christodoulides et al., 2021) in a way that can no longer be ignored by luxury brand managers.

Summary

The present chapter aimed to introduce the reader to the ever-changing world of luxury products and services, presenting the state of the art of academic literature in the field and emerging trends in consumer behavior. First, we reviewed the definition of luxury, describing the unique attributes (e.g., exclusivity, high price, high quality, aesthetic design, etc.)

which characterize luxury goods and services, also focusing on the key role of consumers' perception, highlighting that the concept of luxury is subjective and relative. Then, we discussed the notion of luxury branding, emphasizing the unique "anti-laws" of marketing that differentiate this sector from others. Finally, we briefly described some emerging consumers' needs based on the highlights of the most prominent sectorial studies.

Our primary goal was to emphasize that modern luxury companies must strike a delicate equilibrium between the enduring, unalterable principles that have traditionally defined luxury as we knew it, and the evolving values and demands of contemporary consumers, whether aspirational luxury buyers or extremely rich individuals. For brands to remain relevant over time and ultimately grow their customer bases while fostering loyal customers, they should prioritize innovation, creativity, and responsible initiatives. This may encompass more efficient business processes, applied artificial intelligence for both back-end and front-end interactions with customers, as well as the careful implementation of new business models (e.g., renting and second-hand) which are continuously gaining customer interest. Being at the forefront of sustainability and tech-enabled solutions will remain essential for future competition: while currently we may confidently infer that luxury is inherently sustainable (as we will point out in Chap. 3), we argue that as soon as consumers become more used to it, this may cease to be sufficient. Throughout this book—starting from Chap. 4—we will focus on the key role of aesthetics in shaping responsible customer choices, focusing on some Made in Italy cases (Chap. 5) as an exemplary case of sustainable aesthetic principles.

References

Amatulli, C., & Guido, G. (2012). Externalised vs. internalised consumption of luxury goods: Propositions and implications for luxury retail marketing. *The International Review of Retail, Distribution and Consumer Research, 22*(2), 189–207.

Amatulli, C., De Angelis, M., Costabile, M., & Guido, G. (2017). *Sustainable luxury brands: Evidence from research and implications for managers*. Springer.

American Marketing Association AMA. (2017). Definition of marketing. https://www.ama.org/the-definition-of-marketing-what-is-marketing/

Atkinson, S. D., & Kang, J. (2022). New luxury: Defining and evaluating emerging luxury trends through the lenses of consumption and personal values. *Journal of Product & Brand Management, 31*(3), 377–393.

Bagwell, L. S., & Bernheim, B. D. (1996). Veblen effects in a theory of conspicuous consumption. *The American Economic Review*, 349–373.

Bardhi, F., & Eckhardt, G. M. (2017). Liquid consumption. *Journal of Consumer Research, 44*(3), 582–597.

BCG-Altagamma. (2023). True Luxury Global Consumer Insight 2023. https://www.bcg.com/press/5july2023-true-luxury-global-consumer-insight

Belk, R. W. (1988). Possessions and the extended self. *Journal of Consumer Research, 15*(2), 139–168.

Bellezza, S. (2023). Distance and alternative signals of status: A unifying framework. *Journal of Consumer Research, 50*(2), 322–342.

Berthon, P., Pitt, L., Parent, M., & Berthon, J. P. (2009). Aesthetics and ephemerality: Observing and preserving the luxury brand. *California Management Review, 52*(1), 45–66.

Booms, B. H., & Bitner, M. J. (1981). Marketing strategies and organization structures for service firms. In *Marketing of services* (pp. 47–51). American Marketing Association.

Christodoulides, G., Athwal, N., Boukis, A., & Semaan, R. W. (2021). New forms of luxury consumption in the sharing economy. *Journal of Business Research, 137*, 89–99.

D'Arpizio, C., & Levato, F., (2024) Global luxury spending to land near €1.5 trillion in 2024, remaining relatively flat as consumers prioritize experiences over products amid uncertainty. Bain & Company. Available at https://www.bain.com/about/media-center/press-releases/2024/global-luxury-spending-to-land-near-1.5-trillion-in-2024-remaining-relatively-flat-as-consumers-prioritize-experiences-over-products-amid-uncertainty

D'Arpizio, C., Levato, F., Steiner, A., & de Montgolfier, J. (2024). Long live luxury: Convergence to expand through turbulence, Bain & Company. https://www.bain.com/insights/long-live-luxury-converge-to-expand-through-turbulence/

Dubois, D. (2020). Fulfilling social needs through luxury consumption. *Research Handbook on Luxury Branding*, 75–91.

Dion, D., & Arnould, E. (2011). Retail luxury strategy: Assembling charisma through art and magic. *Journal of Retailing, 87*(4), 502–520.

Dubois, B., Laurent, G., & Czellar, S. (2001). *Consumer rapport to luxury. Analyzing complex and ambivalent attitudes.*

Dubois, B., Czellar, S., & Laurent, G. (2005). Consumer segments based on attitudes toward luxury: Empirical evidence from twenty countries. *Marketing Letters, 16*, 115–128.

Eastman, J. K., Iyer, R., Shepherd, C. D., Heugel, A., & Faulk, D. (2018). Do they shop to stand out or fit in? The luxury fashion purchase intentions of young adults. *Psychology & Marketing, 35*(3), 220–236.

Eckhardt, G. M., & Bardhi, F. (2020). New dynamics of social status and distinction. *Marketing Theory, 20*(1), 85–102.

Eisingerich, A. B., & Bell, S. J. (2008). Perceived service quality and customer trust: Does enhancing customers' service knowledge matter? *Journal of Service Research, 10*(3), 256–268.

Faccioli, G., Martin, K., & Sheehan, E. (2023). Global Powers of Luxury Goods 2023. *Deloitte.* https://www.deloitte.com/global/en/Industries/consumer/analysis/gx-cb-global-powers-of-luxury-goods.html

Foster, R. D. (2024). Exploring 2024 Trends in Global Wealth and The Luxury Market. *Forbes.* https://www.forbes.com/sites/forbes-global-properties/2024/03/14/exploring-2024-trends-in-global-wealth-and-the-luxury-market/?sh=2bc4697d7269

Gong, Y., Zhang, Q., & Zhang, J. (2022). Does renting luxury make me shine? The mediating role of perceived signals. *Psychology & Marketing, 39*(8), 1599–1609.

Guido, G., Amatulli, C., Peluso, A. M., De Matteis, C., Piper, L., & Pino, G. (2020). Measuring internalized versus externalized luxury consumption motivations and consumers' segmentation. *Italian Journal of Marketing, 2020*, 25–47.

Han, Y. J., Nunes, J. C., & Drèze, X. (2010). Signaling status with luxury goods: The role of brand prominence. *Journal of Marketing, 74*(4), 15–30.

Hennigs, N., Wiedmann, K. P., Klarmann, C., Strehlau, S., Godey, B., Pederzoli, D., et al. (2012). What is the value of luxury? A cross-cultural consumer perspective. *Psychology & Marketing, 29*(12), 1018–1034.

Holmqvist, J., Ruiz, C. D., & Peñaloza, L. (2020). Moments of luxury: Hedonic escapism as a luxury experience. *Journal of Business Research, 116*, 503–513.

Ipsos. (2022). Affluent luxury segmentation. https://www.ipsos.com/en-nl/affluent-luxury-segmentation

Kapferer, J. N. (2017). *The end of luxury as we knew it? Advances in luxury Brand management, 25–41.* Springer.

Kapferer, J., & Bastien, V. (2009). The specificity of luxury management: Turning marketing upside down. *Journal of Brand Management, 16*, 311–322. Reprinted in Advances in luxury brand management (2017), 65–84.

Kastanakis, M. N., & Balabanis, G. (2012). Between the mass and the class: Antecedents of the "bandwagon" luxury consumption behavior. *Journal of Business Research, 65*(10), 1399–1407.

Keller, K. (2009). Managing the growth trade-off: Challenges and opportunities in luxury branding. *Journal of Brand Management, 16*, 290–301.

Ki, C. W. C., Li, C., Chenn, A. S., Chong, S. M., & Cho, E. (2024). Wise consumer choices in online secondhand luxury (OSHL) shopping: An integrated model of motivations, attitudes, and purchase intentions for OSHL as wise, conspicuous, and sustainable consumption. *Journal of Retailing and Consumer Services, 76*, 103571.

Ko, E., Costello, J. P., & Taylor, C. R. (2019). What is a luxury brand? A new definition and review of the literature. *Journal of Business Research, 99*, 405–413.

Kumar, A., Paul, J., & Unnithan, A. B. (2020). 'Masstige' marketing: A review, synthesis and research agenda. *Journal of Business Research, 113*, 384–398.

Leibenstein, H. (1950). Bandwagon, snob, and Veblen effects in the theory of consumers' demand. *The Quarterly Journal of Economics, 64*(2), 183–207.

Palombella, I., & Faccioli, G. (2023). Global Powers of Luxury Goods 2023. Game changing steps in luxury. *Deloitte*. https://www2.deloitte.com/it/it/pages/consumer-business/articles/2023-global-powers-of-luxury-goods.html

Pantano, E., & Stylos, N. (2020). The Cinderella moment: Exploring consumers' motivations to engage with renting as collaborative luxury consumption mode. *Psychology & Marketing, 37*(5), 740–753.

Richards, E. A., & Sturman, S. S. (1977). Life-style segmentation in apparel marketing. *Journal of Marketing, 41*(4), 89–91.

Rosendo-Rios, V., & Shukla, P. (2023). The effects of masstige on loss of scarcity and behavioral intentions for traditional luxury consumers. *Journal of Business Research, 156*, 113490.

Scott, J. (2000). Rational choice theory. In Browning, G., Webster, F., & Halcli, A. (2000). Understanding contemporary society: Theories of the present. Understanding Contemporary Society, 126–138.

Silverstein, M. J., & Fiske, N. (2003). Luxury for the masses. *Harvard Business Review, 81*(4), 48–59.

Smith, A. (1776). An Inquiry into the Nature and Causes of the Wealth of Nations. W. Strahan & T. Cadell. Cited in Berthon, P., Pitt, L., Parent, M., & Berthon, J. P. (2009). Aesthetics and ephemerality: Observing and preserving the luxury brand. *California Management Review, 52*(1), 45–66.

Snyder, C. R., & Fromkin, H. L. (1977). Abnormality as a positive characteristic: The development and validation of a scale measuring need for uniqueness. *Journal of Abnormal Psychology, 86*(5), 518.

Tangri, K., & Yu, H. (2023). Why buy used? Motivators and barriers for re-commerce luxury fashion. *International Journal of Retail & Distribution Management, 51*(9/10), 1095–1114.

The Economics. (2023). How the young spend their money. The Economist Group Limited. https://www.economist.com/business/2023/01/16/how-the-young-spend-their-money

Truong, Y., McColl, R., & Kitchen, P. J. (2009). New luxury brand positioning and the emergence of masstige brands. *Journal of Brand Management, 16*(5–6), 375–382.

Vickers, J. S., & Renand, F. (2003). The marketing of luxury goods: An exploratory study–three conceptual dimensions. *The Marketing Review, 3*(4), 459–478.

Vigneron, F., & Johnson, L. W. (1999). A review and a conceptual framework of prestige-seeking consumer behavior. *Academy of Marketing Science Review, 1*(1), 1–15.

Vogel, A. T., Cook, S. C., & Watchravesringkan, K. (2019). Luxury brand dilution: Investigating the impact of renting by Millennials on brand equity. *Journal of Brand Management, 26*, 473–482.

Wiedmann, K. P., Hennigs, N., & Siebels, A. (2009). Value-based segmentation of luxury consumption behavior. *Psychology & Marketing, 26*(7), 625–651.

Wirtz, J., Holmqvist, J., & Fritze, M. P. (2020). Luxury services. *Journal of Service Management, 31*(4), 665–691.

3

Sustainable Luxury and Made in Italy

Luxury companies are currently facing social and economic pressure to adopt sustainable practices, especially from younger generations. The United Nations has defined Sustainable Development as "*development that meets the needs of the present without compromising the ability of future generations to meet their own needs*".[1] Therefore, sustainability encompasses not only the environmental impact of production and consumption choices, but also sustained economic growth, as well as principles of social inclusion and respect for diversity. Even though luxury and sustainability have long been seen as opposing concepts—the former reflecting ostentation, status, and exclusivity, the latter reflecting self-denial, parsimony, and inclusion—this chapter argues that luxury products have some characteristics that make them inherently sustainable: their high quality, durability, limited production, and preservation of artisanal traditions.

[1] United Nation (2016) The Sustainable Development Agenda. What is sustainable development? Available at https://www.un.org/sustainabledevelopment/development-agenda-retired/#:~:text=Sustainable%20development%20has%20been%20defined,to%20meet%20their%20own%20needs.

A. D'Aniello et al., *Aesthetics and Sustainable Luxury Consumption*, Palgrave Advances in Luxury, https://doi.org/10.1007/978-3-031-96833-4_3

The chapter concludes by examining recent research on this subject to discern significant themes and issues associated with contemporary sustainable luxury, particularly with growing business models centered on circular economy and sharing principles, which question some conventional core values of luxury brands.

Luxury and Sustainability

In recent years, consumer awareness around environmental and social issues has grown exponentially, prompting the luxury sector to rethink its role in sustainable development (Amatulli et al., 2018; De Angelis et al., 2017). Once criticized for wasteful production practices—such as the destruction of unsold goods, excessive resource use, and a lack of supply chain transparency—the industry now finds itself under increasing scrutiny from both regulatory bodies and socially conscious consumers (Kapferer & Michaut-Denizeau, 2019). Hereafter, we briefly review literature about luxury and sustainability, focusing on apparent paradoxes and emerging approaches to consumption.

Apparent Contradiction Between Sustainability and Luxury

At first glance, luxury and sustainability may seem two conflicting ideas (Beckham & Voyer, 2014; Janssen et al., 2014; Kapferer & Michaut, 2014; Torelli et al., 2012): they are at best weakly connected (Achabou & Dekhili, 2013), and in some cases they have even been seen as contradictory, or even oxymoronic (Carranza et al., 2023; Hindley et al., 2023). The reason for this is primarily attributed to the common connotations of these terms: luxury is frequently regarded as being synonymous with excess, ostentation, superficiality, hedonism and pleasure-seeking, self-gratification, exclusivity, while sustainability brings to mind notions of self-sacrifice, thriftiness, moderation, self-control, frugality, self-restraint, and inclusivity (e.g., Amatulli et al., 2017; Carrier & Luetchford, 2022; Athwal et al., 2019; Moraes et al., 2017; Nash et al., 2016). To illustrate,

Beckham and Voyer (2014) argued that consumers found it easier to implicitly associate luxury with unsustainability. Simultaneously, it is not arduous to envision that finding a balance between the necessity of inclusivity and the significance of exclusivity poses a challenge for both brands and consumers. Past research suggests that the level of contradiction between sustainability and luxury is reduced for customers who perceive luxury as being of very high quality, but it is significantly higher for those who perceive luxury as being expensive or unusual (Kapferer & Michaut, 2014).

According to Kapferer and Michaut (2014), major criticisms about the sustainability potential of luxury products and brands ground on various aspects of the supply chain, including: (1) the sourcing of raw materials (e.g., animal skins, gold, trees, gemstones), (2) the treatment of animals (e.g., the controversial practice of duck feeding for foie gras, or the killing of young seals for furs; Rolling et al., 2021); (3) the working conditions of employees; and (4) the manufacturing methods that contribute to pollution in local environments.

Moreover, advocates of "unsustainable" luxury also highlight the resulting imbalance of economic and social factors (Amatulli et al., 2017). Indeed, critics of the luxury sector have scrutinized both its social and economic aspects, particularly questioning the rationale behind luxury purchases: they contend that it is illogical to spend a substantial amount of money on a product when an alternative with equivalent capabilities can be acquired on the market at a significantly reduced cost. Hence, luxury has sometimes been blamed for worsening income inequality, especially in emerging countries where extreme poverty coexists with extreme wealth. The luxury consumer is not regarded as innocent but rather accused of deliberately endorsing social problems such as child labor and animal cruelty they associate with luxury firms' operations. In line with this perspective, it is particularly hard for luxury brands to embrace the values of diversity and inclusion if one of the core characteristics of luxury is exclusivity. In some cases, luxury fashion firms have faced criticism for their marketing mistakes, which have included the use of racist iconography and cultural appropriation (Zanotti, 2024), that is a disrespectful and stereotypical way to exploit the elements of a specific culture. For example, in 2018, the Italian fashion brand Dolce & Gabbana was forced to cancel a major fashion show in China after being accused of racism because of a series of videos that showed some stereotypes of Chinese people (Chung & Holland, 2018). Luxury brands are aware of the potential damage to their reputation, especially in light of contemporary communication media, such as social media and blogs, that enable activists, as well as ordinary consumers, to quickly and extensively disseminate rumors (Kapferer, 2010) and negative word-of-mouth (WOM; Amatulli et al., 2020).

Luxury Consumer Behavior: From a Conspicuous to a Consciousness Consumption Approach

The unsustainable spending behavior of the wealthiest customers is a major reason why sustainability appears to be so disconnected from luxury. Laville (2023) reported that the environmental impact of the world's twelve richest billionaires—in terms of global greenhouse gas emissions due to their houses, boats, private jets, and investments—is higher than the yearly energy demand of 2 million average-sized houses. Indeed, the wealthiest individuals usually reside in extremely exclusive properties, frequently with guarded access, and some of them even possess their own private islands (Kenner, 2015). They utilize exclusive modes of transportation such as private aircraft, huge luxury boats, and even submersibles. Ahlström et al. (2020) even found that being able to easily access consumer credit is an additional driver of unsustainable behaviour. Paddison (2021) reports results from a study according to which Bill Gates, who is globally known as being one of the most impactful environmental promoters, traveled on 59 flights only in 2017, for approximately 343,500 km: this resulted in the emission of over 1600 tones of greenhouse gases, which is comparable to the average annual emissions of 105 Americans.

According to a recent BCG-Altagamma, True-Luxury Consumers Insight (2023), the richest customers make up less than 1% of the luxury market, yet they contribute to about 10% of the total worldwide luxury expenditure. Therefore, the substantial purchasing power of these individuals might foster values such as egoism, excessive consumerism, and irresponsibility that are the opposite as those typically associated with sustainability (Beckham & Voyer, 2014). As pointed out by Kenner (2015), one of the biggest challenges for policy makers is that as a result of their considerable wealth, high-net-worth individuals may have the ability to evade the repercussions of climate change and biodiversity decline in their everyday life, thus they may not perceive the pressing need to modify their consumption patterns. This has serious implications in terms of overall consumption trends: Halina Szejnwald Brown, Professor Emerita of Environmental Science and Policy at Clark University, highlights that "*Wealthy people set the tone on consumption to which everybody aspires*" (Paddison, 2021). To illustrate, think about the positioning pay-off of the fast-fast fashion Chinese company "Temu" that invites people to "*shop like a billionaire*",[2] given the very low price of their products which incentivize overconsumption and unnecessary purchases.

While multiple academic studies consistently demonstrate that sustainability is not a relevant factor the majority of luxury consumers consider when making a purchase (Davies et al., 2012; Joy et al., 2012; Kapferer & Michaut, 2014), the transition from "*conspicuous consumption*" (e.g., Achabou & Dekhili, 2013) to "*conscientious consumption*" (e.g., Ki et al., 2017) has triggered consumers from all social classes to become deeply concerned about potential social and environmental issues related to their choices. Moreover, socio-cultural and economic trends have been shifting the motivations behind luxury consumption toward sustainability principles. The contemporary luxury markets are now highly diversified mainly because of greater geographic coverage, increase in global wealth, and generational turnover. Developing economies (mainly China and the Middle East) and younger generations (i.e., Millennials and Generation Z) have seen their discretionary income rise—and so has their relevance to the sector's development (BCG-Altagamma, 2023). By 2030, the Chinese market for luxury goods should overtake the Western economies, reaching the 24%-26% of total global purchases, while Gen X (25%-30%) and Millennials (50%-55%) will dominate the market (D'Arpizio et al., 2024). As a consequence, recent research efforts have mainly been focusing on understanding the

[2] Temu website: https://www.temu.com.

peculiarities of these new luxury buyers: for example, Osburg et al. (2024) examined how sustainability shapes consumers' perceptions of luxury (vs. mass market) product value (in terms of quality, price, emotional, and social benefits) as well as their behavioral intentions (i.e., purchase intention and intention to provide electronic word-of-mouth) in emerging (vs. developed) countries. The authors found that the positive effect of sustainability expressed in terms of higher intention to buy and e-WOM, as well as better product evaluations, was stronger for mass market products in developed countries. Nevertheless, the impact of sustainability on consumers' intentions to purchase luxury products was found to be consistent across different markets, but the effect was stronger in developed markets. Instead, focusing on generations, Pencarelli et al., (2019) aimed to comprehend the desires, behaviors, and purchasing habits of young consumers regarding sustainability, with a particular focus on luxury goods. Their results indicate that corporate social responsibility (CSR) had a noteworthy and favorable influence on the acquisition of luxury goods among both Generation Z and Millennials.

Indeed, these new luxury customers are highly informed, sophisticated, and demanding: they require higher product quality, but they also expect luxury brands to be deeply committed to environmental and ethical issues (Amatulli et al., 2017), preferring sustainable products that align with their personal values and beliefs (Hennigs et al., 2012). Considering that luxury products are primarily used to communicate consumer identity and personal values, it becomes even more crucial for luxury customers that luxury brands take the lead in establishing sustainability standards (Kapferer & Michaut-Denizeau, 2019; Kapferer & Michaut, 2015). Moreover, according to Costly signaling theory (Berger, 2017), there exists a link between altruistic acts and status; thus, choosing a product for environmental protection might serve as a way for individuals to demonstrate their ability to incur significant expenses, thus leading them to prefer green products over conventional luxury ones, especially when shopping in public (Griskevicius et al., 2010). Similarly, the distinction between internalized and externalized luxury consumption (Amatulli and Guido, 2012; Guido et al., 2020) drives different responses in terms of sustainable choices. To illustrate, buyers who purchase luxury products to satisfy their inherent desire for social standing will view sustainability as a less significant driver of choice, unless they perceive it as a something that allows them to be seen (Griskevicius et al., 2010). Conversely, customers who have an internalized approach to luxury tend to adhere to their personal beliefs, with sustainability often being a key factor. As a result, these customers view sustainability in luxury products as something that aligns well with their personality (Amatulli et al., 2017).

In order to sell sustainable luxury products, firms should first deeply understand consumers' purchase motivations and then design effective marketing strategies based on consumers' approaches to luxury consumption. Atkinson and Kang (2022) propose a taxonomy of consumption values that allows to conceptualize the "new luxury": attention capital, inconspicuous consumption, intrinsic experiential value, life enrichment, self-directed pleasure, and, finally, sustainability. Through the application of a cluster analysis on 706 US consumers, Vanhamme et al. (2021) identified five segments of ethical luxury buyers, which differ in terms of levels of consumption of luxury products (i.e., how much they spend in luxury) or ethical products in general (i.e., how much they are interested in buying sustainable products from the mass market), as well as ethical luxury (i.e., how much they are interested in buying sustainable luxury; *see* Table 3.1).

Moreover, a discriminant analysis allowed to identify four different drivers of ethical luxury consumption which can be described as follows (Vanhamme et al., 2021):

(1) *Immorality, ego orientation, and strain*: This dimension reflects the desire to consume luxury goods for the purpose of gaining status as an influential figure (i.e., opinion leader), establishing a positive identity, feelings of dissatisfaction with oneself and a low sense of self-worth, and psychological stress caused by a lack of peace in life. This function is also associated with the low ethical standards of luxury buyers, as indicated by their tendency to report that engaging in illegal activity for personal gain is acceptable. Ethical luxury purchases are driven by factors such as social status, as well as a disrupted worldview characterized by a lack of peace and psychological tension, which are rooted in the egoic realm of the mind (Reisz, 2018).

Table 3.1 Clusters of ethical luxury buyers

Level of luxury consumption		Ethical	Ethical luxury	Characteristics
Light luxury buyers	C1	Light	Light	Second-oldest cluster. Not altruistic-oriented, but also not anxious, immoral, neither driven by egoistic motivations. Not informed nor caring about ethical issues.
	C2	Heavy	Light	Older, not luxury savvy. Not driven by egocentric motives (e.g. being opinion leaders, identity)
Moderate luxury buyers	C3	Moderate	Moderate	Negative perception of people and the world, not feeling in control of positive environmental outcomes. They are also not altruistic-oriented.
Heavy luxury buyers	C4	Heavy	Heavy	Altruistic-oriented (i.e. they want to make their contribution for a sustainable world), young, and savvy luxury consumers
	C5	Heaviest	Heaviest	Extremely ambivalent: they consume ethical luxury because they want to be seen or perceived as opinion leaders (egoistic motivations), they desire to contribute to sustainability (altruistic-oriented), they are young and luxury savvy (informed about luxury and able to differentiate ethical luxury).

Source: Vanhamme et al. (2021)

(2) *Altruistic orientation*: Purchase motivations are determined by elements that suggest good societal connections to ethical luxury, such as the desire to make a positive effect and a concern for ethical issues. From this perspective, the act of indulging in high-quality luxury goods is not driven by ego or a desire to fill a void but rather serves as a way to contribute to a greater good, devoid of selfish motives (Reisz, 2018). Highly informed individuals are being educated about the concept of ethical luxury, which encompasses the notions of creating a positive impact and showing concern for relevant others.

(3) *Conservation, in control, and positivity*: This dimension pertains to individuals who exhibit conservative traits such as conformity and respect for traditions. They also possess a highly optimistic view of the benevolence of the worlds, and their own potential contribution to a better environment (i.e. high self-control and feelings of empowerment).

(4) *Youth and luxury savvy*: The fourth dimension relates to age (exhibiting a negative correlation of aging with intention to engage in sustainable luxury consumption) and a favorable disposition towards luxury. This function tends to differentiate between younger and older luxury buyers, as well as those who possess knowledge and understanding of luxury values and benefits.

To sum up, these results confirm that the desire to actively engage in ethical luxury consumption mostly arises from a range of socio-psychological factors (Dubois et al., 2021), which include biological values (i.e. ethicality and age), socio-psychological factors (i.e. human values, personality, and motives), and structural factors (i.e., assumption about the world) that jointly shapes individuals' perceptions, attitudes, intentions, and behaviors (Vanhamme et al., 2021). Indeed, it is hard to imagine an "ethical consumer" who is completely different from a "traditional consumer" (Carrigan & Attalla, 2001), but there is an increasing awareness of ethical issues that can motivate people to "act ethically" more often in their consumer choices, whether they are acting sincerely, absolving their moral duties, or simply adapting, submitting to social pressures.

The Inherent Sustainability of Luxury

Recent scientific marketing research and empirical evidence support the compatibility between luxury brands and sustainability. This finding aligns with the ongoing efforts of luxury brands to have a good impact on societies and the environment (Amatulli et al., 2017).

Indeed, luxury brands are ready to capitalize on the benefits of sustainability due to their core characteristics, and marketing imperatives, as described in Chap. 2, that make them *inherently* sustainable (Amatulli et al., 2017; De Angelis et al., 2021). To illustrate:

(1) **Quality**: this means selecting high-quality materials and prioritizing meticulous manufacturing processes, which ultimately lead to extended product durability.

    To ensure the sourcing of prestigious raw materials, luxury firms are collectively committed to ensuring a sustainable utilization of these resources in order to safeguard their long-term production capabilities. Moreover, utilizing substandard materials and poor workmanship would diminish the luxury industry's ability to market their products as exclusive. Therefore, it is crucial for luxury brands to allocate resources towards protecting the specialized workforce, infrastructure, and environment that are the basis of their manufacturing quality.

(2) **Scarcity**: this translates into limited production and distribution, which in turn is a way to minimize consumption and strengthen the rarity and exclusivity of their offer. Due to the practice of producing only a limited number of units, premium brands are able to effectively restrict their use of resources (Janssen et al., 2014; Kapferer, 2010). In that way, luxury companies have the ability to undermine resource depletion, mass manufacturing, excessive consumption, and the disposable society in several ways (Kapferer, 2010).

(3) **Heritage**: this means preserving artisanal traditions and history, as well as creating unique iconic products that have the potential to become a source of value which consumers will transfer to the next generations.

Consequently, a growing number of scholars support the idea that luxury is sustainable *by definition*. Indeed, the previously recalled UN definition of sustainable development mainly highlights the concept of "inter-generational equity", which is perfectly compatible with luxury production processes, and durable final outputs: it is not by chance that luxury has been defined as "*the business of lasting worth*" (Amatulli et al., 2017; Kapferer, 2010).

This perspective highlights the environmental sustainability of luxury products, because of their longer durability (Sun et al., 2021), which stems from excellent quality standards, that make them designed to last long over time both from a functional and from a stylistic standpoint (Levinthal & Purohit, 1989).

As pointed out by Sun et al. (2021), the endurance of a product not only reduces waste generation but also provides practical advantages to both customers and companies. Consumers can simultaneously attain sustainability and financial mindfulness by selectively purchasing fewer things (Haws, Winterich, & Naylor 2014). In other words, they can strategically manage their financial budget and contribute to sustainable goals by prolonging the lifespan of their purchases by selective use, reselling, or giving. Indeed, selling a pre-owned bag not only prolongs its lifespan and gives an economic benefit to the seller, but also contributes to environmental preservation.Furthermore, product endurance can also provide advantages for firms. It is a timely characteristic that aligns well with the growing trends of sustainable luxury and "slow fashion" that have gained popularity in recent years (Cline, 2016). Several premium and luxury brands like Patagonia actively encourage the use of top-notch, long-lasting materials that minimize negative environmental effects. Emphasizing the longevity of the product can be an attractive approach for high-end brands to promote both the luxuriousness and the eco-friendliness of their goods (Sun et al., 2021).

Nevertheless, as we previously highlighted, sustainability not only encompasses the environmental impact of production and consumption choices, but also sustained economic growth, as well as principles of social inclusion and respect for diversity. In the next section we will summarize what luxury brands can do to proactively contribute to a comprehensive sustainable development.

Luxury Brands Initiatives for Sustainable Development

While many luxury brand managers have long acknowledged the importance of incorporating sustainability into their operations, many others have neglected this aspect for a long time due to the belief that sustainability was not necessary for achieving success in the luxury market. However, with the proliferation of CSR actions, most brands have made great steps toward sustainability (Amatulli et al., 2017; Amatulli et al.,

2018). CSR is characterized by four dimensions: legal, ethical, economic, and philanthropic (Carroll, 1979). Similarly, John Elkington's (1997) very popular framework, the so-called "Triple Bottom Line" (TBL), expands firms' attention beyond Profit (economic goals) to the careful consideration of their impact on the Planet (environmental goals) and on People (social development goals). Specifically:

- **Planet (ecological dimension)**. The environmental area pertains to a company's engagement with the entire environment, encompassing its utilization of natural resources and its broader influence on the Earth's ecosystems. As we will more thoroughly discuss in Chap. 4, luxury encompasses various distinct industries, each with its own specific focus on environmental conservation. For example, the beauty industry heavily relies on local water supplies and waste management systems, while companies in the leather goods and jewelry sectors, especially those employing exotic leathers, prioritize the conservation of scarce natural resources.
- **People (social dimension)**. Caring about people means carefully managing the potentially conflicting needs of different stakeholders, such as suppliers, consumers, employees, and communities. This means managing issues of workers' conditions throughout the supply chain, taxation, employment standards, human rights, charity, and volunteerism (Carcano, 2013). Company-community relationships work best and are strengthened when both sides understand and accept how they count on and contribute to the other. It would not be fair for a luxury company to treat its workers well if all it did was take from them—steal resources, use ability, and so on—and not give anything back.
- **Profit (economic dimension)**. Economic sustainability means that a business needs to think about "big picture things" if it wants to stay in business for a long time. Companies that follow "sustainability criteria" do better in the business world (Wagner, 2005; Wagner et al., 2001). These outcomes come from having a better governance framework and working together with stakeholders in a more constructive way. Being economically sustainable is important because it is the basis for sustainable growth. Furthermore, ensuring long-term goals and

financially supporting projects that may be good for society and the environment are both possible through economic sustainability.

Below (alongside Fig. 3.1) we briefly summarize the substantial environmental and societal benefits that luxury enterprises can provide for all relevant stakeholders:

1. *Customers* can enjoy superior-quality, long-lasting products that are made from safe and reliable materials and rooted in a "Made In" tradition.
2. *Society* will benefit from greater environmental preservation due to limited production volumes, as well as local and territorial development arising from higher incomes and reputational advancement.
3. *Employees* are treated fairly and receive a secure job with above-average compensation, as luxury brands are particularly resilient to economic downturns. Moreover, firms are incentivized to offer continuous job training in order to preserve and enhance traditional artisan skills and expertise.

More than ever, successful luxury brands are nowadays those that can satisfy the needs of these stakeholders while balancing the accessibility and exclusivity of their offerings. The TBL model provides a unique framework for luxury enterprises to analyze these behaviors and explore alternative options.

Fig. 3.1 How luxury brands are inherently sustainable. (Source: Reproduced from Amatulli et al. (2017))

Sustainability and the "Made-in" Effect

The positive relationship between luxury and sustainability is strengthened when both these concepts are connected with a positive "country of origin" (COO) or "Made In" effect (Amatulli et al., 2017). Indeed, one core characteristic of luxury products is their traditional and artisanal history, which is often reinforced through a "*Made In*" label. The latter is a sign on a product or its packaging, which identifies its geographical origin (Glover & Higgins, 2023), that is usually adopted to emphasize special features of a product because it comes from a specific place (Dalli, 2023). In this section, we argue that many of the values that define sustainability within luxury brand management (shown in Fig. 1) are also fundamental elements that identify the "Made in Italy" brand, thus further emphasizing the inherent sustainability of Italian luxury brands.

Country of Origin and the "Made in" Effect

One characteristic of luxury products is their traditional and artisanal history, which is often reinforced through a "*Made In*" label. Essentially, there is evidence of a country-of-origin effect when customers evaluate two items that are similar in all aspects except for the nation listed on the "Made In" label (Schooler, 1965; Verlegh & Steenkamp, 1999). But what is the relationship between the impact of COO-effect and the concept of "Made In" when it comes to sustainability?

In general, people are inclined to purchase a product and even willing to pay more for a product if it is manufactured in a country renowned for its expertise in that particular industry, that means that the COO-effect is category-specific and strongly related to the cultural specificity of each country. Indeed, there exists strong correlation between the cultural specificities of the various countries and the characteristics of managerial and marketing processes, which in turn affect performances (Ancarani et al., 2020). As a consequence, consumers may associate certain regions with superior quality, while others with poor quality or unethical practices (Laroche et al., 2005). French perfume, Italian clothing, German engineering, and Chinese silk are some examples of favorable cultural links between industries and countries. They function as highly diagnostic information pieces in consumer decision making and indicate the source of the product, as well as the skill, materials, excellence, and longevity associated with it. Therefore, made-in labels are information cues, which potentially affect different stages of consumer behavior from information search and valuation of alternatives to the purchase stage (Samiee et al., 2016).

In other words, COO offers high credibility based on know-how or legitimacy and thus it acts as a brand, conveying intangible values that set all their legitimate products apart from any copy. France means elegance, Italy means aesthetics and “dolce vita”, the United Kingdom means aristocracy, and the United States means wealth and power (Kapferer, 2017).

In other words, they can be powerful means to exploit positive secondary brand associations (Keller, 2013). The “Made In” label, in particular, is considered a crucial factor in ensuring product quality and is an integral part of brand value. The inherent characteristics that render a product both luxury and sustainable (namely, quality and durability, as previously noted) are further enhanced when these attributes are favorably linked to the product’s COO. These connections not only provide luxury companies with a chance to showcase their existing sustainability efforts but also offer opportunities for them to invest in their communities and talent, thereby preserving their heritage and improving the sustainability of their products and business (Amatulli et al., 2017).

Associations with country of origin differ globally. Not all nations have unequivocally favorable COO linkages, and even those that do may discover that their COO associations are advantageous in one industry but have limited impact in another. Germany is renowned for its exceptional engineering skills, exemplified by prestigious automobile brands such as BMW Porsche or Mercedes. Of course, this reputation may not have a significant impact, say, for fashion designers or food producers.

Roth and Romeo (1992) identify four potential COO outcomes:

- The country is viewed favorably, and in a manner that aligns with the product’s key attributes.
- The country is viewed favorably, but in a manner that does not align with the product’s key attributes.

- The country is viewed unfavorably in a manner that aligns with the product's key attributes (i.e., the association with the country of origin would be detrimental).
- The country is viewed unfavorably in a manner that does not align with the product's key attributes.

Every scenario calls for a distinct marketing strategy. When a country's perceived qualities align with the significant features of a product, a favorable match emerges. This should encourage a corporation to emphasize the product's origin. Several possibilities are available, such as:

- Selecting a brand name that accurately represents and emphasizes the image of the country.
- Exploiting design and aesthetics of products and packaging to easily communicate the product's origin through colors and symbols.
- Developing advertising campaigns that explicitly align a country's image with the product category.

Sometimes, the country's perceived strengths do not align with the necessary features of the product. In this scenario, marketing should:

- Emphasize the country-of-origin image as a supplementary advantage.
- Adjust product's attributes to the country's strengths, by choosing brand elements that also exploit positive association with the country (e.g., manufacturing techniques for leather in Tuscany).

When a country's image is negative and the perceived weaknesses align with the relevant attributes of the products (resulting in a negative effect), companies have several options:

- Reduce the connection between the product and its country of origin by emphasizing other product characteristics, such as design or innovativeness.
- Emphasize the aspect(s) in which the product excels, if applicable.
- Adopt a different country's image.

- Collaborate with companies from countries with positive images and utilize the newer designations "Designed in X" or "Made in Y" as much as possible.

When the perceived deficiencies of a country do not align with the relevant attributes of a product (thus neutralizing the COO impact), enterprises should nonetheless exclude any mention of their place of origin, if it is legally acceptable to do so.

Made in Italy and Luxury Values

The "Made in Italy" brand is well recognized and esteemed globally and it has often been recognized as one of the most successful examples of COO branding, as it is among the most well-known brands around the world. Italy benefits from favorable connotations with the country, since the term "Made in Italy" is often seen as linked with creativity, heritage, elegance, innovation, style, elegance, taste, culture (Amatulli et al., 2017). Similarly, according to Temperini et al. (2016) the made in Italy brand evokes quality, creativity, and aesthetics, making Italian productions desirable worldwide.

Italian professors Fabio Ancarani, Michele Costabile, and Marco Francesco Mazzù (2020) conducted a survey on 100 Italian managers to identify six unique strategies of Italian brands to strengthen their global competitiveness and they identified the following elements of "Italian way", as follows:

- **Meticulous attention to detail**: The ability to make attention to every single detail of Italian productions.
- **Adaptive flexibility**: The ability to re-combine technologies based on customers' need for personalization and uniqueness; then derive the development of projects that make it possible to add new and continuous technical and "relational" details to goods, services, and experiences.
- **Thinkering**: The experimental approach that puts doing before designing. It consists in moving towards experimental approaches that make the origin of innovations pragmatic.

- **Creative deconstruction**: The personalization process carried out with unconventional approaches (open innovation). It involves crafting the uniqueness of the product starting from the uniqueness of the entrepreneurial, technological, and commercial processes.
- **Humanistic content**: A core, distinctive element of Made in Italy, the origin of its value proposition. It consists in including the world of art and culture.
- **Sense and aesthetic sensitivity**: The intrinsic and extrinsic excellence of the manufacturing of goods, services, and contents. The main goal is to maintain leadership in design (aesthetic, ergonomic, and functional).

From the consumers' point of view, the two things that mainly characterize the distinctive charm of Italian productions are *coolness* and *uniqueness*, which perfectly align with values that luxury consumers look for when making any consumption choice. Specifically:

- *Coolness*: There exist ten distinct attributes that contribute to a product or brand being perceived as cool, which are: deviation from societal standards, originality, authenticity, the ability to become an element of a subculture, extraordinariness, aesthetic appeal, energy understood as the capacity for emotional activation, "high" status, popularity, and iconicity (Warren et al., 2019). Costabile et al. (2020) propose that all the main distinctive elements of Made in Italy, both in the consumer and industrial markets, contribute to determining the natural coolness of Made in Italy products and services.
- *Uniqueness:* The quest for *uniqueness* is characterized by the search for distinction through the purchase, use, and continuous search for goods and services capable of satisfying the need to avoid similarities and homologations but also to feel and appear unconventional, better if unique. The uniqueness of the products that enjoy the symbolic assurance of made in Italy fulfills the role of helping the consumer to define his identity towards himself, increasing his self-esteem, and towards his own community of reference, increasing their social esteem (Costabile & Mazzù, 2020).

However, not all products can draw equal benefit from the "Made in" effect: In Italy, the product categories that benefit the most are the so-called "3F" (Food, Fashion, Furniture) or "4A" of Made in Italy: Abbigliamento (fashion); Arredamento (furniture); Automazione (machinery); and Agroalimentare (food). Despite economic and political uncertainty, the most important Italian companies and brands continue to grow. Interbrand states that the Italian luxury brands experiencing the most growth in 2024 are those capable of evolving their relationships with customers, such as Ferrari and Prada.[3]

The BREMBO case: Across design, technology, and sustainability
Brembo is a world leader in the design, development, and production of braking systems and components for cars, motorbikes, industrial vehicles, and machinery. In 2020 Brembo won the "ADI Compasso D'Oro" confirming its ability to bring together a distinctive design at the service of functionality and performance, as well as a sense of aesthetics and a love of beauty. Thanks to its extraordinary and original market orientation, Brembo has been able to become a successful example of ingredient branding especially for consumers' evaluation of cars and motorbikes.

Brembo considers environmental preservation to be an integral part of its commitment to achieving optimal performance. Thus, GREENANCE was created as a novel range of groundbreaking products with the objective of minimizing environmental harm. It proactively adheres to the regulations of progressively stringent legislation while ensuring exceptional braking performance. GREENANCE is a brand itself that focuses on creating high-performance and environmentally friendly car components. One of the latest product lines, the GREENANCE Kits, consists of special alloy discs and brake pads. These kits are designed to improve both the performance and sustainability of vehicles. They achieve this by reducing PM10 emissions and PM 2.5 particles, as well as increasing the lifespan of the discs by up to three times.

The GREENANCE Kit, developed through Brembo's extensive research and development, utilizes advanced technological expertise to

[3] Interbrand – Best Global Brands 2024 cited in Forbes (2024) https://forbes.it/2024/10/10/corre-ferrari-debutta-nvidia-100-brand-valgono-piu-mondo/.

achieve the highest levels of technical and quality standards. As a result, it promises a decrease in emissions of over 80%.

Source: Brembo website https://www.brembo.com/it

While striving to enhance Italy's positive and consistent reputation worldwide, the values and associations that characterize "Made in Italy" also carry the potential danger of copying, misuse, or counterfeiting, commonly referred to as "Italian sounding". The latter involves the deceptive utilization of geographical indications, denominations, images, trademarks, and brands to evoke and exploit positive brands association with Italy, in order to promote and sell products that are not actually produced there. The prevalence of Italian sounding marketing constitutes consumer fraud, since companies manipulate consumers into associating Italian characteristics with their products, so obtaining a potential advantage over their competitors, even if those competitors' products are actually manufactured in Italy. As to the European Community Regulation n. 450/2008, the label "Made in Italy" can only be used if the product has been completely manufactured in Italy or if it has undergone a significant transformation within any Italian region. The relevance of the issue is undeniable: just within the food industry, the global worth of counterfeit Made in Italy food has reached 120 billion euros, with the United States being the country with the biggest turnover of counterfeit Italian products.

Emerging Themes in Sustainable Luxury Consumption

While luxury has historically been overlooked by those who are concerned with sustainability (Bendell & Kleanthous, 2007), luxury organizations and enterprises have actively undertaken a consistent process of ecological transformation throughout the entire value chain (Kapferer & Michaut-Denizeau, 2019). This includes sourcing raw materials, production, logistics, HR management, retailing, and post-purchase recycling (Winston, 2016). Similarly, marketing scholars have largely overlooked the sustainable aspect of luxury consumption and have only recently begun to focus their attention on this topic.

In order to encourage more research in this domain, the following sections highlight two key research themes that underlie current academic and managerial debates. In particular, we shed light on: (1) innovative sustainable solutions and (2) emerging business models.

Innovative Sustainable Solutions

Luxury firms have two main avenues for promoting sustainable innovation: (1) adopting more efficient production processes and supply chain management strategies, in order to minimize pollution, waste, energy consumption, and harm to the local territory, and (2) launching new sustainable products through green product design. However, the extant literature lacks a clear understanding of how these approaches to sustainability can benefit or undermine brand performance.

Literature about production processes and supply chain management for sustainable luxury mainly deals with traceability of materials (e.g., Jaegler & Goessling, 2020; Karaosman et al., 2020). Furthermore, it is crucial to efficiently manage key concepts of social innovation, such as diversity and inclusion, as well as employee-centric endeavors (Sipilä et al., 2021), throughout the entire supply chain.

Consequently, transparency and supplier engagement are identified as the most crucial themes for responsible supply chain development (Karaosman et al., 2020). However, luxury brand managers should be aware of the potential detrimental effects of responsible initiatives: while external CSR (e.g., philanthropic support for social causes) is more visible to final customers and can increase their willingness to buy luxury products (Amatulli, De Angelis, et al., 2018), it could lower financial performance and undermine loyalty in the long run, because of customers' perceptions of extrinsic firm motivations (e.g., profit increase; Sipilä et al., 2021). Thus, companies should carefully find the right balance between internal and external initiatives and consider framing their brands as sustainable rather than exclusive (Sipilä et al., 2021). At the same time, finding the right communication strategy is a key priority for both practitioners and scholars (Amatulli et al., 2017; Kong et al., 2021).

Concerning green product design, academic research is still limited and provides conflicting findings about consumers' evaluations of sustainable luxury products (Sani-Elia et al., 2023). According to Kapferer and Michaut (2015), luxury buyers' preference for quality and prestige may be at odds with more sustainable products that seek to use recycled or less resource-intensive materials. Indeed, consumers often perceive sustainable luxury products as lower in quality: Beckham and Voyer (2014) found that only labeling a product as "sustainable" makes luxury items less appealing to consumers. Regarding material innovation, Achabou and Dekhili (2013) found that luxury shoppers disliked the idea of buying a Hermès item made from recycled cotton, while Streit and Davies (2013) found that organic raw materials are detrimental due to customer perceptions of them as "dirty". Similarly, Dekhili et al. (2019) contend that the integration of sustainable materials in luxury goods reduces consumers' perception of luxuriousness, concluding that "*sustainable luxury products […] are deemed incompatible with prestige, self-pleasure, and satisfaction*" (p. 495). These findings indicate that sustainable luxury faces significant challenges in influencing consumers' perceptions of value (Osburg et al., 2021). However, literature reporting positive consumer reactions is growing: Rolling and Sadachar (2018) reported that the use of recycled materials does not decrease millennials' perceptions of luxury; Adıgüzel and Donato (2021) showed that consumers prefer upcycled luxury products over recycled or non-sustainable ones because of high feelings of pride; Nash et al. (2016) found that in the pearl jewelry sector the use of sustainable materials is well-valued by consumers, leading to positive word of mouth.

More generally, further investigation is required to comprehensively comprehend the pivotal significance of aesthetic design for both products and services. Sani-Elia et al. (2023) suggest that consumers respond more favorably to completely new green luxury products than to re-designed existing products because of higher perceived brand sincerity. De Angelis et al. (2017) found that consumers favor new green luxury products when they stylistically match the brand's other offerings compared to when they imitate the models manufactured by non-luxury green companies. These results also highlight the importance of products' aesthetic appearance. In this vein, more research is needed to understand wealthy consumers' shift toward so-called "ugly luxury" (Bellezza, 2023) and its implications for sustainability. Indeed, according to Janssen et al. (2017), sustainable luxury branding is more effective for discreet/inconspicuous luxury rather than for "loud", status-driven items (Han et al., 2010). Therefore, it is possible that "ugly" products will go out of style more easily and prove to be unsustainable, compared to aesthetically pleasing products. The field of green luxury product design partially anticipates what we are going to address in further details in the next chapters: the key role of products and services aesthetics in driving more responsible consumption choices.

Emerging Business Models

As anticipated in Chap. 2, high-status consumers are decreasing their expenditures on material possessions to stand out from mainstream consumers. Frequently, this trend is driven by the desire to adopt a more sustainable lifestyle while avoiding abundance and constant desire for consumption which are typical of materialism (Wilson & Bellezza, 2022). This desire for sustainability may translate in sustainable actions such as owning luxury products for longer time, purchasing pre-loved items, consider renting over buying (e.g., Christodoulides et al., 2021), as well as favoring luxury experiences over product possessions (Atwal & Williams, 2017; D'Arpizio et al., 2021; Kumar, 2024).

The principles of circular economy and the growth of the sharing economy have implications for the luxury market. Indeed, second-hand sales and product rentals are emerging as new models for accessing and owning luxury items (BCG-Altagamma, 2023; Christodoulides et al., 2021). On one side, aspirational consumers have the opportunity to access high-quality products at more affordable prices (e.g., Kumar, 2024). On the other side, habitual luxury buyers are increasingly recognizing the economic and environmental value of reselling (Sun et al., 2021). Bain & Co indicates that the value of the second-hand luxury goods market amounts to €45 billion (nearly 6% of the total luxury expenditures), with developed economies primarily driving this trend: Europe accounts for nearly 50% of second-hand luxury sales, followed by the US (D'Arpizio et al., 2024).

Academic research on the second-hand luxury market has mainly focused on understanding consumers' attitudes and motivations. Turunen and Leipämaa-Leskinen (2015) found that second-hand luxury consumers still derive symbolic value from their products, even without the prospect of exclusivity. Likewise, Amatulli, Pino, et al. (2018) found that vintage luxury products provide their buyers with instrumental (e.g., social approval) and terminal (e.g., self-identification, self-confidence, and self-fulfillment) value (see also Kessous & Valette Florence, 2019). Turunen et al. (2020) adopted the sellers' perspective: the luxury item is scaled down to the driver of a financial transaction, thus the seller feels empowered (change in symbolic value); moreover, they can experience a new sense of social status, or a feeling of being sustainable (change in social role). Recently, Ki et al. (2024) developed an integrated model of Online Second-Hand Luxury (OSHL), which argues that product determinants (i.e., affordability, value retaining, craftsmanship and care), consumer values (e.g., aesthetic taste, lifestyle, casual usage), and online platform characteristics (i.e., authenticity verification and treasure hunting pleasure) shape consumers' perceptions of OSHL as wise, conspicuous, and sustainable consumption, which ultimately drive repurchase intention.

In terms of luxury rentals, Pantano and Stylos (2020) confirm that consumers are more willing to participate in this behavior for five main reasons: (a) the need to find an outfit for a special event; (b) inspiration emerging from recommended brands; (c) willingness to try a new way to exploit a luxury good; (d) the need for greener choices; and (e) longer product life-cycles. Granted, not all consumers who desire unaffordable luxuries are willing to rent them (Gong et al., 2022); thus, luxury brands should carefully consider how they implement rental offerings, given the high risk of decreasing their own perceived brand equity (Vogel et al., 2019).

Beyond being environmentally sustainable, these emerging business models are also enhancing the accessibility of luxury items, which is a crucial aspect of inclusion and diversity. In this realm, an emerging topic that should get more attention in the future is how to balance exclusivity and inclusivity for both product design, new form of consumption, and service environments (e.g., Debenedetti et al., 2024).

Conclusion

New business models, emerging in response to societal shifts and environmental harms, are challenging the traditional tenets of luxury (Christodoulides et al., 2021), but luxury brands can no longer ignore their responsibility towards the environment, the society, and global economic growth at large. Instead, they must find the right balance between their products' unique values and old and new consumers' needs. We envision a continuous improvement process, driven by more mature markets and more experienced customers, that will progressively build appreciation for, and awareness of, the inherently sustainable characteristics of luxury.

Summary

The present chapter aimed to introduce the reader to the key challenges of luxury brands when embracing the values of environmental, social, and economic sustainability. First, we reviewed the reason why sustainability and luxury have often been defined as oxymoronic, describing major reason for criticism in this sector. Therefore, we focused on unsustainable consumer behaviors of wealthiest consumers, but also on the major reasons for a shift towards conscientious consumption which is now driving consumers' choices in a highly diversified market. Then, we briefly explained the reason why luxury and sustainability are indeed strictly related and compatible, focusing on some core characteristics of luxury products (i.e., quality, rarity and heritage), which make them inherently sustainable and timeless (i.e. long-lasting). Besides the concept of product durability, we further highlighted other key aspects of sustainability: namely, sustained economic growth, and social responsibility. Furthermore, we emphasized the key role of "Made in" and "country of origin" in strengthening the relationship between luxury brands and sustainability, focusing on the exemplary case of Made in Italy as a way to leverage positive brand association with a country and, at the same time, highlighting sustainable aspect of brands. Indeed, luxury and sustainability can go hand in hand, as shown by the "Made In" phenomenon. It is, in fact, one way for high-end brands to strengthen their own environmentally friendly strategies. Luxury brands that value craftsmanship and culture also value the power of scarcity in their products. This is because they use manufacturing methods and know-how that are usually associated with "Made In" production values, such as a focus on quality, the preservation of artisans, support for the local supply chain, and protection of raw materials. All of these things may help the products last a long time and give the brand a long-term vision. Finally, we discussed some emerging trends and future research directions that require scholars' attention in the future. Our primary goal was to emphasize that modern luxury companies cannot ignore sustainable development objectives. Starting from Chap. 4, our work will advance a consumer-based view of sustainable luxury, built around the idea that it is consumers' experiences with the product that shape length of product usage, above and beyond products' *tangible* durability. Specifically, grounding on the idea that a key component of the luxury product experience is its aesthetics (Cesareo et al., 2023; Kapferer and Valette-Florence, 2021), we will propose that certain aesthetics characteristics of luxury products might affect consumers' choice, use, and disposal of products, thus potentially offering a positive contribution to sustainable consumption.

References

Achabou, M. A., & Dekhili, S. (2013). Luxury and sustainable development: Is there a match? *Journal of Business Research, 66*(10), 1896–1903.

Adıgüzel, F., & Donato, C. (2021). Proud to be sustainable: Upcycled versus recycled luxury products. *Journal of Business Research, 130*, 137–146.

Ahlström, R., Gärling, T., & Thøgersen, J. (2020). Affluence and unsustainable consumption levels: The role of consumer credit. *Cleaner and Responsible Consumption, 1*, 100003.

Amatulli, C., De Angelis, M., Costabile, M., & Guido, G. (2017). *Sustainable luxury brands: Evidence from research and implications for managers*. Springer.

Amatulli, C., De Angelis, M., Korschun, D., & Romani, S. (2018). Consumers' perceptions of luxury brands' CSR initiatives: An investigation of the role of status and conspicuous consumption. *Journal of Cleaner Production, 194*, 277–287.

Amatulli, C., De Angelis, M., Pino, G., & Guido, G. (2020). An investigation of unsustainable luxury: How guilt drives negative word-of-mouth. *International Journal of Research in Marketing, 37*(4), 821–836.

Amatulli, C. & Guido, G. (2012). Externalised vs. internalised consumption of luxury goods: Propositions and implications for luxury retail marketing. *The International Review of Retail, Distribution and Consumer Research, 22*(2), 189–207.

Amatulli, C., Pino, G., De Angelis, M., & Cascio, R. (2018). Understanding purchase determinants of luxury vintage products. *Psychology & Marketing, 35*(8), 616–624.

Ancarani, F., Costabile, M., & Mazzù, M. F. (2020). Specificità culturali e qualità delle produzioni. Economia & management: la rivista della Scuola di Direzione Aziendale dell'Università L. *Bocconi, 2*, 40–43.

Athwal, N., Wells, V. K., Carrigan, M., & Henninger, C. E. (2019). Sustainable luxury marketing: A synthesis and research agenda. *International Journal of Management Reviews, 21*(4), 405–426.

Atkinson, S. D., & Kang, J. (2022). New luxury: Defining and evaluating emerging luxury trends through the lenses of consumption and personal values. *Journal of Product & Brand Management, 31*(3), 377–393.

Atwal, G., & Williams, A. (2017). Luxury brand marketing–the experience is everything!. Advances in luxury brand management, 43–57.

BCG-Altagamma (2023). True Luxury Global Consumer Insight 2023. Available at https://www.bcg.com/press/5july2023-true-luxury-global-consumer-insight

Beckham, D., & Voyer, B. G. (2014). Can sustainability be luxurious? A mixed-method investigation of implicit and explicit attitudes towards sustainable luxury consumption. *ACR North American Advances, 42*, 245–250.

Bellezza, S. (2023). Distance and alternative signals of status: A unifying framework. *Journal of Consumer Research, 50*(2), 322–342.

Bendell, J., & Kleanthous, A. (2007). Deeper luxury. WWF-UK, Godalming, 9–15.

Berger, J. (2017). Are luxury brand labels and "green" labels costly signals of social status? An extended replication. *PloS one, 12*(2), e0170216.

Carcano, L. (2013). Strategic management and sustainability in luxury companies: The IWC case. *Journal of Corporate Citizenship, 52*, 36–54.

Carranza, R., Zollo, L., Díaz, E., & Faraoni, M. (2023). Solving the luxury fashion and sustainable development "oxymoron": A cross-cultural analysis of green luxury consumption enablers and disablers. *Business Strategy and the Environment, 32*(4), 2399–2419.

Carrier, J. G., & Luetchford, P. G. (Eds.). (2022). *Ethical consumption: Social value and economic practice*. Berghahn Books.

Carrigan, M., & Attalla, A. (2001). The myth of the ethical consumer–do ethics matter in purchase behaviour?. *Journal of Consumer Marketing, 18*(7), 560–578.

Carroll, A. B. (1979). A three-dimensional conceptual model of corporate social performance. *Academy of Management Review, 4*, 497–505.

Cesareo, L., Townsend, C., & Pavlov, E. (2023). Hideous but worth it: Distinctive ugliness as a signal of luxury. *Journal of the Academy of Marketing Science, 51*(3), 636–657.

Cho, R. (2021). *Why fashion needs to be more sustainable*. Columbia Climate School. https://news.climate.columbia.edu/2021/06/10/why-fashion-needs-to-be-more-sustainable/

Christodoulides, G., Athwal, N., Boukis, A., & Semaan, R. W. (2021). New forms of luxury consumption in the sharing economy. *Journal of Business Research, 137*, 89–99.

Chung, S., & Holland, O. (2018). Dolce & Gabbana cancels China show amid 'racist' ad controversy. *CNN*. https://edition.cnn.com/style/article/dolce-gabbana-shanghai-controversy/index.html

Costabile, M., Fei, C., & Mazzù, M. F. (2020). Perché e come difendere la nostra coolness. Economia & management: la rivista della Scuola di Direzione Aziendale dell'Università L. *Bocconi, 2*, 44–48.

Costabile, M., & Mazzù, M. F. (2020). La domanda e l'offerta di unicità. Economia & management: la rivista della Scuola di Direzione Aziendale dell'Università L. *Bocconi, 2*, 49–52.

D'Arpizio, C., Levato, F., Prete, F., Gault, C., & de Montgolfier, J. (2021). The future of luxury: Bouncing back from Covid-19. Available at https://www.bain.com/insights/the-future-of-luxury-bouncing-back-from-covid-19/

D'Arpizio, C., Levato, F., Steiner, A., & de Montgolfier, J. (2024). *Long live luxury: Convergence to expand through turbolence*. Bain & Company. https://www.bain.com/insights/long-live-luxury-converge-to-expand-through-turbulence/

Dalli, D. (2023). *Marketing perspectives on the country of origin in National Brands and Global Markets* (pp. 24–45). Routledge.

Davies, I. A., Lee, Z., & Ahonkai, I. (2012). Do consumers care about ethical luxury? *Journal of Business Ethics, 106*(1), 37–51.

De Angelis, M., Adıgüzel, F., & Amatulli, C. (2017). The role of design similarity in consumers' evaluation of new green products: An investigation of luxury fashion brands. *Journal of Cleaner Production, 141*, 1515–1527.

De Angelis, M., Amatulli, C., & Petralito, S. (2021). Luxury and sustainability: An experimental investigation concerning the diamond industry. *Sustainable Luxury and Jewelry*, 179–198.

Debenedetti, A., Philippe, D., & Dion, D. (2024). Balancing exclusivity and inclusivity through the strategic domestication of the luxury retail experience. *International Journal of Research in Marketing, 42*, 32–54.

Dekhili, S., Achabou, M. A., & Alharbi, F. (2019). Could sustainability improve the promotion of luxury products? *European Business Review, 31*(4), 488–511.

Donato, C., De Angelis, M., & Amatulli, C. (2020). Sustainable luxury: The effect of corporate social responsibility strategy on luxury consumption motivations. In *Understanding luxury fashion: From emotions to brand building* (pp. 123–146). Springer Verlag.

Dubois, D., Jung, S., Ordabayeva, N. (2021). The psychology of luxury consumption. *Current Opinion in Psychology, 39*, 82–87.

Elkington, J. (1997). The triple bottom line. *Environmental management: Readings and cases, 2*, 49–66.

Glover, N., & Higgins, D. M. (Eds.). (2023). *National brands and global markets: An historical perspective*. Taylor & Francis.

Gong, Y., Zhang, Q., & Zhang, J. (2022). Does renting luxury make me shine? The mediating role of perceived signals. *Psychology & Marketing, 39*(8), 1599–1609.

Griskevicius, V., Tybur, J. M., & Van den Bergh, B. (2010). Going green to be seen: Status, reputation, and conspicuous conservation. *Journal of Personality and Social Psychology, 98*(3), 392.

Guido, G., Amatulli, C., Peluso, A. M., De Matteis, C., Piper, L., & Pino, G. (2020). Measuring internalized versus externalized luxury consumption motivations and consumers' segmentation. *Italian Journal of Marketing, 2020*, 25–47.

Han, Y. J., Nunes, J. C., & Drèze, X. (2010). Signaling status with luxury goods: The role of brand prominence. *Journal of Marketing, 74*(4), 15–30.

Haws, K. L., Winterich, K. P., & Naylor, R. W. (2014). Seeing the world through GREEN-tinted glasses: Green consumption values and responses to environmentally friendly products. *Journal of Consumer Psychology, 24*(3), 336–354.

Hennigs, N., Wiedmann, K. P., Klarmann, C., Strehlau, S., Godey, B., Pederzoli, D., … & Oh, H. (2012). What is the value of luxury? A cross-cultural consumer perspective. *Psychology & Marketing, 29*(12), 1018–1034.

Hindley, C., van Stiphout, J., & Legrand, W. (2023). Luxury hospitality and sustainability: An oxymoron or viable pursuit? In J. S. Chen (Ed.), *Advances in hospitality and leisure* (Vol. 19, pp. 3–23). Emerald Publishing Limited.

Jaegler, A., & Goessling, T. (2020). Sustainability concerns in luxury supply chains: European brand strategies and French consumer expectations. *Business Strategy and the Environment, 29*(6), 2715–2733.

Janssen, C., Vanhamme, J., & Leblanc, S. (2017). Should luxury brands say it out loud? Brand conspicuousness and consumer perceptions of responsible luxury. *Journal of Business Research, 77*, 167–174.

Janssen, C., Vanhamme, J., Lindgreen, A., & Lefebvre, C. (2014). The Catch-22 of responsible luxury: Effects of luxury product characteristics on consumers' perception of fit with corporate social responsibility. *Journal of Business Ethics, 119*, 45–57.

Kapferer, J. N. (2010). All that glitters is not green: The challenge of sustainable luxury. *European Business Review, 2*(4), 40–45.

Kapferer, J. N. (2017). The end of luxury as we knew it? *Advances in Luxury Brand management*, 25–41.

Kapferer, J. N., & Michaut, A. (2014). Is luxury compatible with sustainability? Luxury consumers' viewpoint. *Journal of Brand Management, 21*(1), 1–22.

Kapferer, J. N., & Michaut, A. (2015). Luxury and sustainability: A common future? The match depends on how consumers define luxury. *Luxury Research Journal, 1*(1), 3–17.

Kapferer, J. N., & Michaut-Denizeau, A. (2019). Are millennials really more sensitive to sustainable luxury? A cross-generational international comparison of sustainability consciousness when buying luxury. *Journal of Brand Management*, 1–13.

Kapferer, J. N., & Valette-Florence, P. (2021). Which consumers believe luxury must be expensive and why? A cross-cultural comparison of motivations. *Journal of Business Research, 132*, 301–313.

Karaosman, H., Perry, P., Brun, A., & Morales-Alonso, G. (2020). Behind the runway: Extending sustainability in luxury fashion supply chains. *Journal of Business Research, 117*, 652–663.

Keller, K. L. (2013). *Strategic brand management: Building, measuring, and managing brand equity*. Pearson Education Limited.

Kenner, D. (2015). Inequality of overconsumption: The ecological footprint of the richest. *Global Sustainability Institute*, 1–18.

Kessous, A., & Valette-Florence, P. (2019). "From Prada to Nada": Consumers and their luxury products: A contrast between second-hand and first-hand luxury products. *Journal of Business Research, 102*, 313–327.

Ki, C., Lee, K., & Kim, Y. K. (2017). Pleasure and guilt: How do they interplay in luxury consumption? *European Journal of Marketing, 51*(4), 722–747.

Ki, C. W. C., Li, C., Chenn, A. S., Chong, S. M., & Cho, E. (2024). Wise consumer choices in online secondhand luxury (OSHL) shopping: An integrated model of motivations, attitudes, and purchase intentions for OSHL as wise, conspicuous, and sustainable consumption. *Journal of Retailing and Consumer Services, 76*, 103571.

Kong, H. M., Witmaier, A., & Ko, E. (2021). Sustainability and social media communication: How consumers respond to marketing efforts of luxury and non-luxury fashion brands. *Journal of Business Research, 131*, 640–651.

Kumar, J. (2024). Psychological mechanisms behind access-based luxury brand consumption: Empirical investigation from the lens of new ownership paradigm. *Journal of Business Research, 175*, 114539.

Laroche, M., Papadopoulos, N., Heslop, L. A., & Mourali, M. (2005). The influence of country image structure on consumer evaluations of foreign products. *International Marketing Review, 22*(1), 96–115.

Laville, S. (2023). Twelve billionaires' climate emissions outpollute 2.1m homes, analysis finds. *The Guardian*. https://www.theguardian.com/environment/2023/nov/20/twelve-billionaires-climate-emissions-jeff-bezos-bill-gates-elon-musk-carbon-divide

Levinthal D. A., Purohit D. (1989). Durable goods and product obsolescence. *Marketing Science, 8*(1), 35–56.

Moraes, C., Carrigan, M., Bosangit, C., Ferreira, C., & McGrath, M. (2017). Understanding ethical luxury consumption through practice theories: A study of fine jewellery purchases. *Journal of Business Ethics, 145*(3), 525–543.

Nash, J., Ginger, C., & Cartier, L. (2016). The sustainable luxury contradiction: Evidence from a consumer study of marine-cultured pearl jewellery. *Journal of Corporate Citizenship, 63*, 73–95.

Osburg, V.-S., Davies, I., Yoganathan, V., & McLeay, F. (2021). Perspectives, opportunities and tensions in ethical and sustainable luxury; Introduction to the thematic symposium. *Journal of Business Ethics, 169*(2), 201–210.

Osburg, V. S., Yoganathan, V., Bartsch, F., Diallo, M. F., & Liu, H. (2024). How sustainable luxury influences product value perceptions and behavioral intentions: A comparative study of emerging vs. developed markets. *Journal of Business Ethics, 191*(4), 713–738.

Paddison, L. (2021). How the rich are driving climate change. *BBC.* https://www.bbc.com/future/article/20211025-climate-how-to-make-the-rich-pay-for-their-carbon-emissions

Pantano, E., & Stylos, N. (2020). The Cinderella moment: Exploring consumers' motivations to engage with renting as collaborative luxury consumption mode. *Psychology & Marketing, 37*(5), 740–753.

Pencarelli, T., Ali Taha, V., Škerháková, V., Valentiny, T., & Fedorko, R. (2019). Luxury products and sustainability issues from the perspective of young Italian consumers. *Sustainability, 12*(1), 245.

Reisz, J. (2018). *The new luxury: How rising global consciousness is killing the old luxury business.* CreateSpace Independent Publishing Platform.

Rolling, V., & Sadachar, A. (2018). Are sustainable luxury goods a paradox for millennials? *Social Responsibility Journal, 14*(4), 802–815.

Rolling, V., Seifert, C., Chattaraman, V., & Sadachar, A. (2021). Pro-environmental millennial consumers' responses to the fur conundrum of luxury brands. *International Journal of Consumer Studies, 45*(3), 350–363.

Roth, M. S., & Romeo, J. B. (1992). Matching product catgeory and country image perceptions: A framework for managing country-of-origin effects. *Journal of International Business Studies, 23*(3), 477–497.

Samiee, S., Leonidou, L. C., Aykol, B., Stöttinger, B., & Christodoulides, P. (2016). Fifty years of empirical research on country-of-origin effects on consumer behavior: A meta-analysis. In *Rediscovering the essentiality of marketing: Proceedings of the 2015 academy of marketing science (AMS) world marketing congress* (pp. 505–510). Springer International Publishing.

Sani-Elia, I., Perez, D., & Grinstein, A. (2023). Born to be sustainable: Consumers' response toward luxury products that are born versus reborn sustainable. *Journal of the Association for Consumer Research, 8*(3), 327–338.

Schooler, R. D. (1965). Product bias in the Central American common market. *Journal of Marketing Research, 2*(4), 394–397.

Sipilä, J., Alavi, S., Edinger-Schons, L. M., Dörfer, S., & Schmitz, C. (2021). Corporate social responsibility in luxury contexts: Potential pitfalls and how to overcome them. *Journal of the Academy of Marketing Science, 49*, 280–303.

Streit, C. M., & Davies, I. A. (2013). Sustainability isn't sexy. Sustainability in fashion and textiles: Values, design, production and consumption, 207.

Sun, J. J., Bellezza, S., & Paharia, N. (2021). Buy less, buy luxury: Understanding and overcoming product durability neglect for sustainable consumption. *Journal of Marketing, 85*(3), 28–43.

Temperini, V., Gregori, G. L., & Palanga, P. (2016). The brand made in Italy: A critical analysis. *Management Studies, 4*(3), 93–103.

Torelli, C. J., Monga, A. B., & Kaikati, A. M. (2012). Doing poorly by doing good: Corporate social responsibility and brand concepts. *Journal of Consumer Research, 38*(5), 948–963.

Turunen, L. L. M., Cervellon, M. C., & Carey, L. D. (2020). Selling second-hand luxury: Empowerment and enactment of social roles. *Journal of Business Research, 116*, 474–481.

Turunen, L. L. M., & Leipämaa-Leskinen, H. (2015). Pre-loved luxury: Identifying the meanings of second-hand luxury possessions. *Journal of Product & Brand Management, 24*(1), 57–65.

Vanhamme, J., Lindgreen, A., & Sarial-Abi, G. (2021). Luxury ethical consumers: Who are they? *Journal of Business Ethics*, 1–34.

Verlegh, P. W., & Steenkamp, J. B. E. (1999). A review and meta-analysis of country-of-origin research. *Journal of Economic Psychology, 20*(5), 521–546.

Vogel, A. T., Cook, S. C., & Watchravesringkan, K. (2019). Luxury brand dilution: Investigating the impact of renting by Millennials on brand equity. *Journal of Brand Management, 26*, 473–482.

Wagner, M. (2005). How to reconcile environmental and economic performance to improve corporate sustainability: corporate environmental strategies in the European paper industry. *Journal of Environmental Management, 76*(2), 105–118.

Wagner, M., Schaltegger, S., & Wehrmeyer, W. (2001). The relationship between the environmental and economic performance of firms: what does theory propose and what does empirical evidence tell us?. *Greener Management International, 34*, 95–108.

Warren, C., Batra, R., Loureiro, S. M. C., & Bagozzi, R. P. (2019). Brand coolness. *Journal of Marketing, 83*(5), 36–56.

Wilson, A. V., & Bellezza, S. (2022). Consumer minimalism. *Journal of Consumer Research, 48*(5), 796–816.

Winston, A. (2016). Luxury brands can no longer ignore sustainability. Harvard Business Review, pp. 2–4.

Zanotti, B. (2024). *Balancing the paradox of inclusivity and exclusivity: How luxury fashion brands use communications to address diversity and inclusion.* Doctoral dissertation, Toronto Metropolitan University.

4

Luxury Aesthetics

Aesthetics have long been a cornerstone of marketing, serving as a powerful tool to shape consumer perceptions, drive engagement, and establish brand identity (Hagtvedt, 2023). Marketing strategies, including product and service design choices about functionality and aesthetics, are formulated to attract and retain consumers, fulfilling their social and personal objectives (Charters, 2006; Levy & Czepiel, 1974). While primarily associated with art, aesthetics goes beyond basic visual appearance of an object or a space, encompassing the entire domain in which individuals experience interest, pleasure, and emotion in relation to the presence or absence of beauty (Levy & Czepiel, 1974, p. 87). Research suggests that aesthetic appeal and aesthetic experiences shape consumers' cognitive and emotional responses, which in turn affect the way they evaluate products, brands, and even themselves (Hagtvedt, 2023). As we briefly anticipated in the previous chapters, the relevance of aesthetics is amplified within the luxury sector where sensory appeal, superior quality, cutting-edge design, aesthetic enjoyment, and craftmanship converge to create memorable experiences (Hagtvedt, 2023; Wang, 2022). Given the pervasiveness of aesthetics in marketing and luxury, a rigorous approach to the field is required. To fill this gap, we believe that marketers can take

A. D'Aniello et al., *Aesthetics and Sustainable Luxury Consumption*, Palgrave Advances in Luxury, https://doi.org/10.1007/978-3-031-96833-4_4

advantage of consumer neuroscience methods and metrics to better understand individual reactions to the use of aesthetics in the marketplace. In particular, the emerging discipline of Neuroaesthetics studies the biological bases of aesthetic experiences by looking at the cognitive and emotional processes related to experiencing art, objects, or the natural environment (Chatterjee & Vartanian, 2014).

Hereafter, we will review the existing literature on aesthetics in marketing, and in particular on luxury. Afterwards, we will provide a detailed review of consumer neuroscience and the way it can contribute to the better understanding of consumers' motivation, decision making, memory, and emotional response to luxury aesthetics.

Aesthetics in Marketing

Aesthetics is ubiquitous in marketing, and it serves multiple functions such as creating value, identifying brands, affecting consumer choices, and ultimately enhancing customer experiences (Hagtvedt, 2023; Patrick et al., 2019). According to the Oxford dictionary,[1] aesthetic is "*the branch of philosophy that studies the principles of beauty, especially in art*". However, providing a universal definition of aesthetics is not an easy task, since it is a multifaceted construct that comes from the intersection of different fields, namely art and design (Hagtvedt, 2023). When it comes to marketing, the relationship between aesthetics and marketing is dual: on one side, marketing can be used to promote work of arts, on the other side strategies about products, packaging, brand identity, and advertising require everyday aesthetic considerations (Levy & Czepiel, 1974). Our work emphasizes the latter, characterizing aesthetics as "*beautiful sensory information*" (Hagtvedt, 2023, p. 8), from which individuals can derive experiences that are meaningful, enjoyable, intriguing, or emotionally evocative (Hagtvedt, 2022a). Although partially overlapping, aesthetics is not synonymous with design: some non-designed objects (e.g., food/fruit/natural environment) can be analyzed in terms of aesthetics;

[1] Definition of aesthetic noun from the Oxford Advanced Learner's Dictionary. Available at https://www.oxfordlearnersdictionaries.com/definition/english/aesthetic_2.

likewise, design also includes considerations of functionality, user-friendliness, and ergonomics (Hagtvedt, 2023). However, as luxury products and service environments typically result from meticulous design considerations, and since design elements also affect aesthetic experiences, throughout the book we will often refer to "aesthetics", "aesthetic design", or "visual design" as synonymous.

Aesthetic Principles

Aesthetic principles refer to the compositional strategies which are used to logically organize the structural elements of a product's form to achieve universal appeal (Coates, 2003). According to Kumar (2015) balance, emphasis, movement, pattern, proportion, harmony, and variety as key aesthetic principles. Hereafter, we present some core principles of aesthetics which have been identified by previous literature as key aesthetic principles: *typicality, novelty, congruity, harmony, unity, symmetry, dynamism, complexity, contrast, and space* (e.g., Hagtvedt, 2023; Li et al., 2013; Wang, 2024).

- *Typicality (or prototypicality)* refers to how well an object represents its category. Research indicates that typical designs initially trigger positive responses from consumers (Landwehr et al., 2013), while atypicality positively influences impressions of brand excitement and negatively influences impressions of brand reliability (Schnurr, 2017). Indeed, as consumers become more familiar with a product's attributes and features, their preference for typical designs wanes, and they start favoring atypical designs that stand out. Heitmann et al. (2020) distinguish between brand typicality (similarity within the brand's range) and segment typicality (similarity within the competitive environment).
- *Novelty* is strictly related to the concept of atypicality, as it refers to the extent to which consumers recognize a product as completely new or different from past experiences, thus recognizing it as new (Li et al., 2013). Consumers may react positively to novelty, both for product design and store displays (Keh et al., 2021), although they may need time to get used to it.

- *Congruity* refers to the state of matching or agreement between elements (Hagtvedt, 2023). Marketing research has explored perceived congruity between brand personality and aesthetic principles in terms of effect for brand equity (Luffarelli et al., 2019) or in the context of logo design (Brasel & Hagtvedt, 2016).
- *Harmony* is the degree to which sensory elements in a design form a coherent, unified pattern (Kumar & Garg, 2010). Similarly, *unity* refers to the cohesive harmony among various design elements. In aesthetic research, beauty is seen as an orderly and harmonious relationship, where the sense of beauty arises from overall harmony (Wang, 2024).
- The antithesis of harmony is the concept of *contrast*, which can manifest across various sensory inputs that characterize product and service experiences, including visual (e.g., shapes, colors), tactile (e.g., smooth versus rough surfaces; Chen et al., 2025), olfactory (e.g., high versus low intensity of fragrances), as well as auditory and gustatory (Hagtvedt, 2023).
- *Symmetry* is often referred to as the simplest case of balance (Creusen et al., 2010), which can be easily identified in objects that display two identical halves that are mirror-inverted (Treder, 2010). Symmetry has been found to convey perceptions of brand premiumness (Pombo & Velasco, 2021), increase purchase intention (Bigoin-Gagnan & Lacoste-Badie, 2018) while asymmetry is associated with higher brand excitement (Bajaj & Bond, 2018).
- *Dynamism* denotes actual or perceived movement that can elicit either positive or negative consumers judgments, contingent upon whether it is deemed engaging or bothering (Hagtvedt, 2023).
- *Complexity* refers to the level of stimulus from the quantity and quality of physical elements, their differences from previous experiences, and the organization of these elements. Classical aesthetic theory suggests that moderate complexity, rather than extreme, induces the highest pleasure (Wang, 2024). Generally, consumers prefer low complexity (Creusen et al., 2010).
- Many of the above-discussed principles deal with the management of *space*, that can be applied in product or logo design (e.g., interstitial space for textual brand logo, Gupta & Hagtvedt, 2021) or even in

managing stores and service environments (e.g., Sevilla & Townsend, 2016). For example, Sevilla and Townsend (2016) discovered that elevated space-to-product ratios in retail displays enhance perceptions of product aesthetic appeal and store prestige. In some other cases, emptiness may not be preferred by customer (Hagtvedt, 2023).

Other relevant aesthetic principles, including *proportion and balance* (associated with quantities such as the number of units, lengths of objects, or the variety of colors and shapes), *ambiguity* (referring to the potential for multiple interpretations of an object or advertisement's aesthetic), *imitation of nature* (Hagtvedt, 2023), or *art injection* (i.e., the integration of artistic elements into products, Li et al., 2013) are also worth mentioning. Further, consumer reactions to aesthetic stimuli are likely to be influenced by individual differences, and other contextual factors, including culture, and current fashion which will be briefly reviewed in the next paragraph.

Aesthetics and Consumer Behavior

Baisya & Das (2008, p. 44) define aesthetics in marketing as "*the study of buyers' cognition, affective and behavioral response to media, entertainment and arts*". Similarly, according to Wang (2024) the aesthetic experience in the consumption process refers to the moment when consumers' aesthetic realm is triggered, induced, and resonated by products or services in the purchasing scene, achieving visual integration and the unity of subject and object, thus highlighting the psychological activity of consumers stimulated by various aesthetic elements (Wang, 2024). In terms of cognition and emotions, Hagtvedt (2023) effectively synthetizes main outcomes of aesthetic stimuli in terms of fluency, interest, pleasure, and arousal. For example, fluency is often associated with typicality and simplicity: while a complex service design diminishes fluency, thereby reducing its attractiveness (Orth & Wirtz, 2014), sometimes complexity can be fluently processed as soon as it is also considered typical (Landwehr et al., 2011). At the same time, novelty may stimulate curiosity and arousal (Buechel & Townsend, 2018), while Kumar and Garg (2010) found harmony to be positively related to pleasure.

Prior research consistently confirmed that aesthetically pleasing goods have the potential to increase sales (Luchs et al., 2016), drive financial decisions (Townsend & Shu, 2010) or intention to donate (Carvalho et al., 2019; Townsend, 2017). Moreover, aesthetics can shape the perception of a brand personality (Bajaj & Bond, 2018) or its premiumness (Pombo & Velasco, 2021), the attitude towards a product (Baghirov & Zhang, 2024), or even the way consumers make inferences about product functionalities (Crolic et al., 2019) and other characteristics such as healthiness (Schnurr, 2017, 2019) and safety (Gupta & Hagtvedt, 2021). Furthermore, given that consumption satisfies both individual and social needs, aesthetics of products and service design also affect the way people perceive the self and others (Hagtvedt, 2023). For example, people may avoid unattractive products because it would negatively affect how they perceive themselves (Grewal et al., 2019), while affirming their sense of self through high aesthetic design (Townsend & Sood, 2012). It's important to mention that aesthetic appreciation is intrinsically subjective, since it strictly depends on individual taste and personality (Hoyer & Stokburger-Sauer, 2012; Wang, 2022). For example, Amatulli et al. (2021) found that perceived atypicality increases consumers' willingness to buy for consumers with high need for uniqueness. Su et al. (2019) found that consumers who view themselves as socially excluded respond more favorably to dense visual patterns. Song et al. (2021) found that consumers with low (vs. high) power-distance beliefs prefer user-designed (vs. designer-designed) products. Jiang et al. (2019) found that consumers with higher financial resources enjoy more angular (vs. rounded) shapes. Furthermore, consumers' perception of beauty results from both objective product attributes and their cultural subjectivity. For example, Li et al. (2013) found that when *visual impressions* align with a consumer cultural values and morals, they are also seen as aesthetically pleasing by that group.

Despite globalization, aesthetic taste varies across *cultural contexts*. For example, white represents purity in Western weddings, while red signifies festivity in Chinese weddings (Wang, 2024). *Fashion* denotes the prevailing lifestyle trends within a certain era and context, originating from modern technology with artistic aesthetics (Wang, 2024). Contemporary fashion signifies that a thing is readily embraced and cherished during a specific timeframe, thus meaning that a product that is out of the current fashion may struggle to meet contemporary customers' needs (Wang, 2024). As pointed out by Hagtvedt (2023), certain cultural disparities may fade over time. Similarly, fashion is inherently dynamic, necessitating an ongoing demand for fresh insights about consumer preferences and aesthetic outcomes.

Aesthetics in Luxury

In Chap. 2 we defined aesthetic appeal and aesthetic value as key luxury brand features, which means that luxury products and services are carefully designed to be pleasant and aesthetically appealing (see section "Aesthetic Dimensions of Luxury"). Indeed, aesthetic appeal is strictly linked to the concept of high quality, since luxury should simultaneously embrace "*aesthetic and technical excellence*" (Dion & Arnould, p. 504). Accordingly, scholars recognize a strict relationship between art, luxury, and aesthetics (De Angelis et al., 2020; Hagtvedt, 2020; Hagtvedt, 2022b; Joy et al., 2014; Kapferer, 2014; Vigneron & Johnson, 2017). In an article originally published in Journal of Brand Management in 1997, Kapferer claimed that "*Luxury defines beauty, it is art applied to functional items*" (Kapferer, 2017, p. 238). Thus, it is not surprising that luxury products are often considered "works of art" due to their beauty and refinement (Venkatesh et al., 2010). Kapferer (2014) posits that linking luxury to art serves as a tactic for businesses to distinguish themselves from growing competitors attempting to replicate luxury codes, hence enhancing the symbolic value of authentic luxury.

Aesthetic Dimensions of Luxury

An ongoing qualitative research related to this book identified seventeen specific dimensions of luxury aesthetics: *visual impact, beauty, attention to detail, style, elegance, recognizability, colors, memorability, originality, minimalism, harmony, shapes, simplicity of lines, craftsmanship, design, quality, and cultural heritage* (De Angelis et al., 2025). These dimensions collectively create a multifaceted construct that elevates luxury goods above ordinary products. Hereafter we introduce these dimensions as pertinent to what distinguishes luxury aesthetics from other categories (then in section "Aesthetic Experiences: A Consumer-Based View of Sustainability in Luxury", we will present the main results of a statistical analysis comparing the effects of different aesthetic elements in terms of sustainable consumer behaviors).

Visual impact is a critical element in luxury aesthetics, as luxury products are designed to create a profound visual experience that triggers emotional responses and enhances value perceptions (Okonkwo, 2016). Respondents viewed luxury aesthetics as an attempt at evoking a rich visual experience. *Beauty* is another fundamental aspect, often serving as the primary allure for luxury consumers (Dion & Arnould, 2011; Kapferer, 2017). Consistently, respondents associated luxury products with something successful, beautiful, and sinuous. The meticulous *attention to detail* found in luxury items reflects the high-quality artisanal production, further distinguishing these products (Kapferer, 2014). Respondents reported that thinking about the aesthetics of luxury makes them reflect on some specific details which makes the product or service recognizable. Additionally, *style* in luxury goods is closely linked to personal pleasure and gratification, driven by both cognitive and emotional responses (Godey et al., 2009; Wang, 2022). Elegance encompasses refinement, distinction, and sophistication, and it is a hallmark of luxury aesthetics (Roux et al., 2017). The *recognizability* of luxury products, often marked by unique design features or logos, also sets them apart (Janssen et al., 2017). The use of specific *color* schemes and combinations enhances the aesthetic appeal of luxury items (Cho & Suh, 2020; Kim et al., 2020). *Originality* is often associated with their rarity, uniqueness, and exclusivity. *Minimalism* is characterized by simplicity and limited decoration, and is often preferred in luxury aesthetics, highlighting the sophistication and elegance of the products (Wilson & Bellezza, 2022). *Harmony* is achieved through a coherent and unified visual composition, which in turn enhances the attractiveness of luxury goods and services (Kim et al., 2006). Similarly, our respondents associated luxury aesthetics with harmony: For example, one respondent noted that "aesthetics should be understood as that which is endowed with outward beauty and yet evokes a feeling, such as harmony". In terms of *shapes*, the *geometric structure* and form of luxury items play a significant role in their aesthetic appeal (Van Rompay & Pruyn, 2011; Wang, 2022). Clean and *simple lines*, often associated with higher aesthetic value, are a distinctive feature of luxury products (Liu et al., 2024). Finally, the emphasis on *quality*, *design*, *craftsmanship*, and *cultural heritage* underscores the handmade quality and traditional manufacturing excellence inherent in luxury goods (Amatulli et al., 2017). Indeed, many of these elements commonly define luxury goods and this was very clear in consumers' minds: respondents linked luxury product design—which is often "purely artisanal" and connected to territory, with the quality. In this regard, as claimed in the previous chapters the "Made in Italy", often described as being *bello e ben fatto* (BBE)[2]—which literally means *beautiful and well-made*—provides a notable example. Some respondents of the qualitative interviews explicitly referred to Italian brands when making examples about luxury aesthetics, such as the harmonious lines of Armani and Bottega Veneta, the unique style of Ferrari, and the quality of leather of Tod's shoes.

Emerging Trends

Prior research suggests that high-quality aesthetics can create pleasant feelings and associations with luxury, shaping consumer perceptions of status, elegance, and higher quality (Shi et al., 2021). Moreover, luxury aesthetics involves a sophisticated sensory experience, characterized by primary sensory elements such as shapes, and spatial structures, along with comprehensive styles and themes (Wang, 2022). This sensory richness contributes to the cultivation of consumers' aesthetic taste, making the interaction with luxury goods deeply engaging and pleasurable. One key contribution to the field of luxury and aesthetics is the one of Berthon et al. (2009) who propose that luxury is not just an object, but rather the process of engaging with it, that can be fully appreciated only if

[2] Confindustria Centro Studi (2023) Esportare la Dolce Vita. The "beautiful and well-made": the potential of made in Italy in the global landscape. Report 2023. Available at https://www.confindustria.it/home/centro-studi/prodotti/EDV/rapporto/tendenze+delle+imprese+e+dei+sistemi+industriali/rapporto-esportare-la-dolce-vita-2023.

individuals have aesthetic knowledge (i.e., the ability to recognize and value exceptional quality).

As we will also illustrate in Chaps. 5 and 6, luxury brand managers can exploit aesthetic principles and features to design products, packaging, stores, advertisement, and service environments. For example, the design and quality of packaging contribute significantly to the overall luxurious experience, enhancing appeal and reinforcing the brand's premium status, and high-end packaging is designed to evoke a sense of anticipation and delight (Romeo-Arroyo et al., 2023). Packaging design in luxury markets is crafted to create a memorable unboxing experience, reinforcing the product's value and the brand's prestige (Romeo-Arroyo et al., 2023).

One of the main challenges for luxury brands is to embrace creativity and innovation, while being coherent with their standard aesthetic design. Recently, research has documented a growing consumer interest toward the so-called "ugly luxury" (Cesareo et al., 2023; Zanette et al., 2022), that is products that challenge common beauty standards and thus represent a unique opportunity for high-end consumers who wish to signal their status by resisting the democratization of luxury among mainstream consumers (Bellezza, 2023). While philosophers, psychologists, and marketing experts agree that there is no universal definition of beauty, ugliness has always been defined as the opposite of beauty (Cohen, 2013). Ugly products are typically evaluated negatively, evoking distaste (Bloch, 1995), dislike, and disgust (Bloch et al., 2002). However, when it comes to luxury the outcome is quite different. And yet, there is a particular ugly aesthetic in luxury fashion, examples including Balenciaga's Triple S sneakers and Gucci's multi-patterned tiger emblazoned sweaters, that is expected to continue being successful in future (Cesareo et al., 2023). According to the fashion critic Alexander Fury, Miuccia Prada is the leading authority of creating unconventional and unattractive products, so that popular press agrees that "ugly chic" can be considered as a distinctive point of this Italian fashion brand, as much as sexiness is associated to Versace. More generally, luxury businesses are looking for new ways to establish a distinctive aesthetics that are attractive to their target customers and that convey a sense of exclusivity and uniqueness. According to Berthon et al. (2009) this requires managing a balance between timeless luxury and ephemeral trends like limited-editions, or crafted experiences.

Consumer Neuroscience

Cognitive neuroscience studies the neural bases of cognition with particular focus on brain activity (Pozharliev et al., 2017). In the past decade the interest in and applications of cognitive neuroscience has grown exponentially driven by the modern brain imaging instruments and the advanced data analysis techniques which enable scientists to associate brain activity with actual behavior to a degree of reliability unimaginable decades ago. Besides, current studies go beyond the clinical field to address diverse manifestations of human behavior. The twentieth century was a determining factor in neuroscience due to the advances made in the knowledge of the nervous system, its mechanisms and, ultimately, the neurobiological processes that determined behavior. In this context, consumer neuroscience is the study of the neural conditions and processes that underlie consumption, their psychological meaning, and their behavioral consequences (Reimann et al., 2011). The birth of consumer neuroscience has generated wide-ranging, ongoing debates on whether this hybrid field benefits its parent disciplines namely consumer psychology and neuroscience (Ariely & Berns, 2010).

To fully understand the role of neuroscience in analyzing and understanding consumer psychology and behavior, and consequently, its role within the aesthetic dimension, it is pivotal to learn more the parts of the brain that are associated with key decision-making processes and the neuroscientific foundations of consumer neuroscience.

The Consumer Brain

The activity of the nervous system is the result of the functioning of billions of cells, organized in circuits that exchange messages, producing perceptions, language, movements, emotions, and thoughts (Michael-Titus and Shortland (2022). The peculiar appearance of the cerebral hemispheres is due to the configuration of the cerebral cortex with its grooves and convolutions. These grooves serve to delimit the four lobes (Fig. 4.1). The central fissure separates the frontal lobe from the parietal; the lateral fissure separates the frontal lobe, the temporal lobe, and the parietal lobe; the occipital lobe is located at the back of the hemispheres.

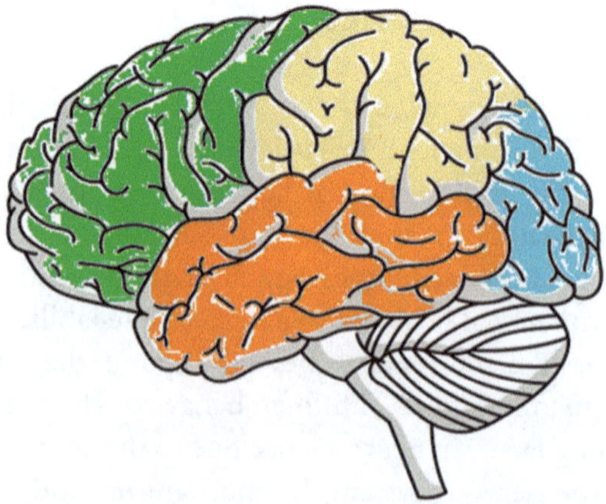

Fig. 4.1 Brain structure. (The frontal (green), temporal (orange), occipital (blue), and parietal lobes (yellow) are distinguished. The brain is seen from the left, being the frontal area on the figure's left. Source: Authors' elaboration)

From an evolutionary perspective, research on the brain development suggests that the human brain is organized into three distinct structures (Fig. 4.2). Compared to the four lobes and Brodmann classifications (Brodmann, 2007), the following one is simpler but intuitive and corresponds to the brain's evolutionary stages and functions. The Triune Brain is a theoretical model of the human brain devolvement proposed by Paul Donald MacLean (MacLean, 1985), an American doctor specializing in neuroscience who made essential contributions to psychiatry.

On the deepest level we find the reptilian or primitive brain (yellow color in Fig. 4.2), its oldest part. It is characterized by quick decisions that prevail over the midbrain (limbic system) or the rational brain (neocortex). Therefore, it consumes 90% of the brain's energy. On the second level we find the Limbic system (red color in Fig. 4.2) is known as the emotional system. It is a structure developed by mammals about 100 million years ago. The limbic system is associated with functions such as procreation, struggle, self-preservation, and, above all, our functional behavior. It comprises different brain structures including the hippocampus or the amygdala that regulate emotions and the consequent physiological responses. The limbic system plays a critical role in learning and is closely related to memory, attention, personality, and behavior.

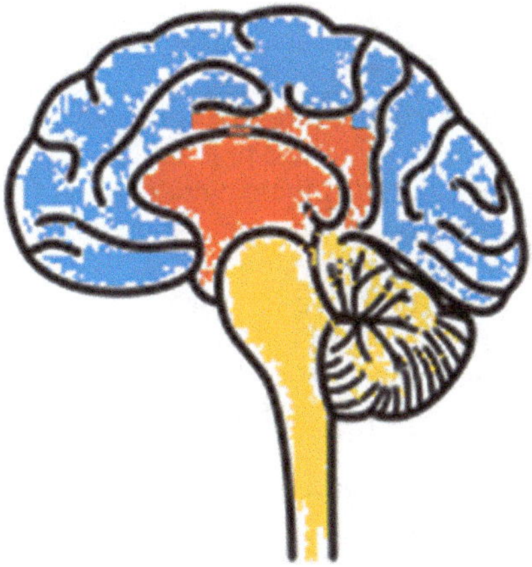

Fig. 4.2 The Triune brain. (Source: Authors' elaboration)

The neocortex (blue color in Fig. 4.2) is the most evolved structure of our brain, and it is what allows humans to differentiate themselves from other animals. It represents 76% of brain volume. It is the place where we find higher cognitive functions, such as language, the ability to reason abstractly, the ability to imagine the future, planned actions, and all forms of conscious experience.

Among these, in the last decades, the limbic system has raised attention for its role in managing emotion. A core theoretical aspect of the limbic system framework is that emotional experiences arise from integrating sensations from the external world with information from the body, specifically feedback from the viscera (Dalgleish, 2004). The notion that emotional experience emerges from feedback from the body in tandem with a higher-level representation of the world partially echoes the seminal theories of emotion put forward by Lange (Lange, 1885). These controversial models argued that emotional experience arises directly from the perception of change in the body.

The use of advanced neuroimaging techniques allows increasingly precise identification of neural responses across specific brain regions (Table 4.1). Such anatomical localization is essential due to the potential relevance of specific brain regions for consumer research, given their role in different cognitive and emotional functions.

Table 4.1 Brain areas and consumer neuroscience

Brain region	Main functions
Amygdala	Processing of emotions, significantly negative emotions; aversive responses to inequity. A modulator of the memory system (Rilling & Sanfey, 2011)
Anterior cingulate	Integration of emotional and motivational information in the decision-making process. Internal conflict between alternative options (Bush et al., 2000; Sanfey et al., 2003)
The dorsolateral prefrontal cortex (DLPFC)	Cognitive control, including over impulses towards social norm compliance (Rilling & Sanfey, 2011)
Hippocampus	Formation and consolidation of memory (long-term memory). Acquisition and recall of declarative memory (McGaugh, 2000; Morris et al., 2024)
Insula	Perception and expectation of risks, pleasure-disgust processing and loss aversion (Sanfey et al., 2003; Wang et al., 2023)
Occipital lobe	Processing of visual stimuli and attention (Armstrong et al., 2006)
Orbitofrontal cortex (OFC)	Evaluation of the capacity of outcomes to satisfy one's needs (Wallis, 2007). Experience and anticipation of the emotion of regret when outcomes differ from expectations.
Ventral Striatum	Evaluation of actual rewards to expectations (O'Doherty, 2004).
The ventrolateral prefrontal cortex (VLPFC)	Representation of the threat of punishment for non-compliance with social norms (Rilling & Sanfey, 2011)
The ventromedial prefrontal cortex (VMPFC)	Processing different alternatives and their perceived value (Daw et al., 2006).
Nucleus accumbens	Behavioral preferences and reward outcomes during the risky decision making. Plays a crucial role in action selection, integrating cognitive and affective information processed by frontal and temporal lobe regions to augment the efficiency and vigor of appetitively or aversively motivated behaviors (Floresco, 2015; Sugam et al., 2012).

Source: Authors' elaboration

Neuroscience and Consumer Psychology

The development of neuroscience research in consumer behavior provides insights for the direct measurement of people's emotional experiences. The possibility of recognizing brain activity during the execution of cognitive processes in humans represents a challenge to our understanding of the relationship between brain and action. It simultaneously leads to a new theoretical approach in various fields, undermining previous ones. The discoveries and methods developed in the neuroscientific field have undoubtedly played an increasingly important role in marketing (Casado-Aranda & Sanchez-Fernandez, 2022).

Consumer neuroscience helps marketers to better understand the relationship between customer behavior and the neurophysiological system. Using knowledge and insights from human brain anatomy and physiological functions of different brain areas, it is possible to model neuronal activity underlying specific human behaviors. Through neuroimaging methods, researchers can compare different brain area activations during a specific task to develop new models which can not only describe the dynamics of human decisions but also understand the usual mismatches between consumers' thoughts and their actions (Pozharliev et al., 2021a). A comprehensive analysis of the discipline can be found in the review article published by Cherubino, Martinez-Levy, and colleagues in 2019 (Cherubino et al., 2019).

Neuroscience Techniques and Key Consumer Processes

According to Kenning and Plassmann (2005) consumer neuroscience tools can be divided into three separate groups: the ones that record metabolic activity, the ones that record electrical activity in the brain, and the ones that do not record electrical activity in the brain (Fig. 4.3).

Through neuroimaging methods, researchers can compare different brain area activations during a specific task to develop new models which can not only describe the dynamics of human decisions but also understand the usual mismatches between consumers' thoughts and their actions (Pozharliev et al., 2017). Each technology has advantages and

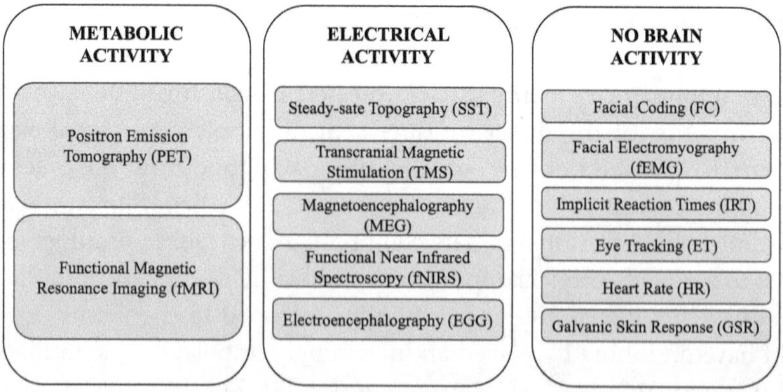

Fig. 4.3 Neuroscience tools used in consumer behavior. (Source: Authors' elaboration)

disadvantages that every experienced marketer needs to consider. Technologies that measure the brain's metabolic activity, such as functional magnetic resonance imaging (fMRI) and positron emission tomography (PET), are relatively more expensive but provide more detailed information about brain activations. Technologies used to measure the brain's electrical activity such as electroencephalography (EEG) don't offer full access to the brain but are less expensive and user friendly (Pozharliev, Rossi, & De Angelis, 2022a). Technologies that measure non-brain activity are highly useful for obtaining physiological data or information on consumers' visual behavior without making the respondents aware of its use (Pozharliev et al., 2022b). For example, facial coding and eye tracking require only a camera to gather data on facial movements and gaze patterns.

These neuroimaging techniques can help marketers measure marketing-related processes such as decision-making, reward processing, memory, attention, and emotional processing through specific brain area activations. A summary of the most common processes related to consumer behavior will be provided in the following section.

Decision Making

Several regions of the prefrontal cortex (PFC) situated in the brain's frontal lobe play a key role in the underlying processes of human decision-making. Past research demonstrates that both the orbitofrontal cortex (OFC) and the ventromedial prefrontal cortex (VMPFC) are involved in decision-making processes. Specifically, these regions are engaged when assessing the perceived value of different options and potential outcomes (Daw et al., 2006). Notably, the OFC is associated with the evaluation of trade-offs and the expected capacity of the outcomes to satisfy one's needs (Wallis, 2007). It plays a central role in choosing appropriate behaviors, especially in unpredictable situations. Measuring these regions' activity can thus provide useful insights into marketing constructs such as perceived value and the neural foundations of consumer choices.

Reward System

The reward is defined as a physiologic event experienced subjectively as pleasure, which increases the likelihood of engaging in acts that in the past were followed by such pleasure (Roese et al., 2017). Different brain regions which respond to subjectively attractive rewards such as food, money, and drugs are involved in the reward system process. Attractive elements like product design or a preferred brand can be considered rewarding stimuli within consumers' brains, which may trigger the psychological motivations that influence purchase behavior. Neuroscience assessment of reward offers a tool for marketers to know more about consumer preferences. For example, when designing a new product, consumer neuroscience can help identify the product elements that are most likely to be well perceived by customers. Further, it can help marketers find the right communication strategy to promote the product. The ultimate objective is to uncover the so-called "gut-level" reward responses which are more likely to predict future market demand (Ariely & Berns, 2010; Reimann et al., 2010). On this behalf, some authors suggest that liking and wanting processes rest on separate neural systems and may sometimes be uncoupled (Dai et al., 2010). For the marketer, this means that consumer neuroscience methods used to measure wanting will help in assessing consumer anticipation of a new product experience, whereas methods used in assessing liking will help in assessing pure enjoyment and hence repeat purchase. In terms of brain regions, the "wanting" system involves a network of neurons that connect two critical areas: the prefrontal cortex and the limbic system. The experience of pleasure, on the other hand, is associated with the "liking" system, which includes pathways extending from the ventral pallidum to the nucleus accumbens (Roese et al., 2017). Customer satisfaction is greater when the actual reward exceeds expectations.

Attention and Memory

Consumers are exposed to an enormous amount of information every day, even though their processing capacity is limited. Each second, people receive 11 million bits of unfiltered information (in the form of advertisements, products, brands, images, colors, sounds) through their five senses. Most of the input goes unnoticed, given that humans can process a small part of it, around 50 bits (Wilson, 2002). For marketers it is important to understand what consumers direct their attention to once exposed to more than one rapidly identified choice alternatives (i.e., brands and products). Paying attention to something, a mechanism of selecting information that gets prioritized over other available information, means a person needs to be aware they are paying attention to it. Four conceptual components are fundamental to attention: bottom-up or saliency filters, top-down control, competitive visual selection, and working memory (Knudsen, 2007). Since aesthetic components are a prominent part of advertisements, logos, and product designs, they are processed by the brain on a visual level and are deconstructed by into their constituent elements such as shapes and colors (Cherubino et al., 2015). Thus, the underlying brain mechanisms of attention and visual processing are of a great interest to consumer research.

Prior research has found that the prefrontal cortex plays a key role in attention and has been linked to the neurons responsible for processing visual stimuli in the occipital lobe (Armstrong et al., 2006). Similarly, studying memory-related mental processes might provide helpful insights into variables influencing consumer behavior, such as brand awareness, product experience, and advertising recall. Furthermore, to perform complex cognitive tasks, people must maintain access to large amounts of information. The hippocampus, located in the temporal lobe, plays a key role in generating different forms of memory and memory processing and consolidation (McGaugh, 2000), i.e., in long-term memory, the acquisition and recall of declarative memory (Eichenbaum, 2000). Additionally, the amygdala, situated next to the hippocampus, is an important modulator of the memory system, particularly in memory consolidation (McGaugh, 2000).

Emotional and Motivational Processing

Emotions drive consumer choices and are very important in the decision-making process (e.g., Huang, 2001). The amygdala, a central brain structure, has a pivotal role in the regulation of emotional responses. It is involved in the processing of negative emotions, unknown stimuli, and aversive responses to inequity (Rilling & Sanfey, 2011). It is also known as a locus of aversive and fear memory. Concerning positive emotions, it is involved even if to a minor extent than negative emotions, usually about rewarding stimuli (Maren & Quirk, 2004). Another critical emotion-related region is the insula (or insular cortex) which plays an essential role in the processing of negative experiences such as the perception and expectation of risks, especially when making decisions for which a social or financial risk is expected (Preuschoff et al., 2008).

Further, indirect variables of emotional processing could be gathered by tracking the activity of specific anatomical structures' variations linked to the emotional processing activity in human beings, such as the prefrontal and frontal cortices (Davidson & Irwin, 1999). The structurally and functionally heterogeneous PFC region plays a well-recognized role in generating emotions (Davidson, 2000). FMRI studies have shown that, when a reward is being enjoyed, activity in the orbitofrontal cortex (OFC), particularly in its medial parts, correlates with subjective reports about the pleasantness valence of the experience. The difference between the left and right cortex activity, also known as frontal asymmetry, is a valid indicator of motivational processes, whether positive or negative (Davidson, 2004). Moreover, prior research has found strong relationships between people's behavioral approach system (BAS) and the behavioral inhibition system (BIS) and their consumer-related activities (Pozharliev, Rossi, & De Angelis, 2022c).

Further, it is possible to assess the emotional responses by looking at peoples' biometrics such as the heart rate (HR) and skin conductance (SC). Indeed, emotions are accompanied by (bodily) reactions that are partially beyond an individual's control (Pozharliev et al., 2022b). These autonomic reactions include facial expressions (e.g., smiling and frowning) and physiological reactions (e.g., sweating) primarily caused by changes in the autonomic nervous system (ANS). Several instruments have been developed to capture autonomic reactions over the years, as the validity of self-reports for measuring lower-order emotions is often biased by cognitive or social desirability constraints. So, the measurement of autonomic reactions can overcome the validity problem of self-report because they measure emotional responses with non-invasive techniques beyond the respondents' control.

Consumer Neuroscience Applied to Luxury and Aesthetics

Thus far, we have examined how neuroscience might enhance our comprehension of consumer behavior, by looking at the different areas of the brain and their specific functions. In the following paragraphs we will review the available literature on neuroscience applied to luxury and aesthetics.

Consumer Neuroscience and Luxury

The application of consumer neuroscience to study consumer responses to luxury products and services has been proliferating in the last decade. However, it is still a field in which more research is needed. A recent study in Italy has looked at the brain responses of 16 luxury buyers (Balconi et al., 2019). Participants were divided into two groups according to their level of sensitivity to environmental issues (high vs. low) and stimuli were presented both explicitly (by a salesperson) and implicitly (images). Both groups, regardless of the attitude toward sustainability, show greater beta-band, delta and theta cortical activity in the temporoparietal area (compared to frontal areas), in the condition of implicit stimulation (imagery). The delta and theta brain activities were used as an indicator of an underlying emotional type of process elicited by the observation of emotional stimuli. More specifically, theta-band activity was found to be higher in the right temporoparietal area (lateralization effect), with higher levels for the non-sustainability-oriented group than for the sustainability-oriented group. The presence of theta waves in both groups indicated a negative emotional response experienced by the participants following the observation of the stimulus-images, thus indicating a high level of involvement.

Zhang et al. (2019) explored consumers' implicit motivations for purchasing luxury brands by using event-related potentials (ERPs). They concluded that consumers' preferences for luxury brands are based on the satisfaction of their social goals. These social goals always coexist and perform as compensation with each other. The dissatisfaction of one social goal would promote their expectation of the satisfaction of another social goal.

Audrin et al. (2018) used an eye tracking to investigate how psychological values (i.e., materialism) modulate visual attention to specific sources of information (i.e., product, brand, and additional information) in the context of luxury consumption. They demonstrated that individuals with high levels of materialism may prefer luxury products for different reasons than individuals with low levels of materialism: while the first focuses on the symbolic dimension conveyed by the brand, the latter pays attention to the actual product characteristics (Audrin et al., 2018).

Pozharliev et al. (2015) studied the neurological responses elicited by luxury (vs. basic) branded products. Further, the authors looked at the role of the social context, namely the mere presence effect on the brain responses of luxury consumers. The findings indicate that luxury products evoke stronger emotional responses and sustained attention, as reflected by higher P2 and P3 amplitudes in event-related potentials

(ERPs). The presence of another person amplifies these responses, with significantly higher late positive potential (LPP) amplitudes observed for luxury (vs. basic) items in social settings. This suggests that luxury products are perceived as more emotionally significant, particularly when others are present, enhancing their motivational value.

While these studies provide critical insights into the neural mechanisms underlying luxury consumption, there remains considerable potential for further exploration. Research by Balconi et al. (2019) has demonstrated how luxury stimuli elicit emotional engagement through theta-band activity, while Zhang et al. (2019) have shown that luxury preferences are shaped by the satisfaction of social goals. Additionally, Audrin et al. (2018) have highlighted how materialism influences visual attention in luxury consumption, and Pozharliev et al. (2015) have demonstrated that the presence of others amplifies neural responses to luxury products. The literature review from Zhao et al. (2023) incorporates findings from neuromarketing and consumer psychology that can guide strategies to improve the perception of sustainable luxury. In particular, they suggest that the brain processes sustainability differently depending on the consumer's predisposition; emotions play a key role in the purchase decision, with pride and novelty as positive factors; and that sustainable luxury should be communicated without generating cognitive dissonance, ensuring that the message reinforces the perception of exclusivity and quality (Zhao et al., 2023).

Despite these publications in the luxury sector, much research is still lacking on how luxury consumption is processed at the neural level. By integrating advanced neuroimaging techniques, real-life consumption scenarios, and experimental methodologies, consumer neuroscience can better unravel the implicit cognitive and emotional drivers of luxury consumption and ultimately refine marketing strategies and improve brand-consumer relationships in a scientifically grounded manner.

Neuroscience and Aesthetics

The emergence of neuroimaging technologies in the 90s gave way to new disciplines within cognitive neuroscience such as Neuroaesthetics (Chatterjee & Vartanian, 2014). It represents an interdisciplinary field that examines how human brain perceives beauty and artistic value at the neural level (Salah & Salah, 2008), which draws on the field of empirical aesthetics (Nadal & Vartanian, 2022).

The term Neuroaesthetics was introduced for the first time by Zeki (1999) who studied the connection between art and activations in the visual cortex. The visual brain is characterized by a set of parallel processing-perceptual systems and a temporal hierarchy in visual perception (Zeki, 1998). Thus, the author proposes that there is also modularity, a functional specialization, in visual aesthetics. Zeki (1999) concludes his work by affirming that an approach to the biological foundations of aesthetics is likely to enhance the sense of beauty—the biological beauty of the brain (Zeki, 1999).

Neuroaesthetics, therefore, studies how the brain processes support aesthetic behavior. Although the real experimental work on this issue has only begun in the last 20 years or so, the conceptual inclination to investigate the neural mechanisms underlying aesthetic behavior can be traced back to the eighteenth and nineteenth centuries (Skov et al., 2017). Neuroaesthetic literature is mainly focused on aesthetic appraisal, that is directly asking participants to judge a visual stimulus as beautiful or ugly. For example, Kawabata and Zeki (2004) employed fMRI to study the brain correlates while participants evaluated paintings (e.g., landscapes, portraits) as beautiful, neutral, or ugly. The medial orbitofrontal cortex (OFC) activated differently for judged-beautiful stimuli than neutral and ugly stimuli (Kawabata & Zeki, 2004). The differential activation observed in the OFC consisted of decreased activity concerning baseline, with judged-ugly stimuli evoking the lowest level of activation. Boccia et al. (2016) investigated the neural correlates of visual aesthetic experience (VAE) and found that the interaction between the Executive Control Network (ECN) and the Default Mode Network (DMN) results in top-down and bottom-up processor activation. On this behalf, Cattaneo et al. (2015) discovered that aesthetic appreciation is related to the activation of areas of the frontal cortex—example: left dorsolateral prefrontal cortex (DLPFC) and the posterior parietal cortex; with visual cortical areas—example: lateral occipital cortex.

Although aesthetic appreciation and aesthetic judgment of beauty use different cognitive processes, these neurological findings suggest that they may recruit similar brain areas that facilitate emotion. Taken together, the aesthetic experience may depute the activation of neural networks and their associated nodes engendering reward computation and emotional processes, which may contribute to both the viewer's judgement of beauty and their appreciation of visual artwork, which is also affected by emotions (e.g., Pelowski et al., 2017).

Research suggests that aesthetic experiences engage two primary neural processes (Menninghaus et al., 2019): (1) the aesthetic evaluation of beauty, based on perceptual judgments, and (2) aesthetic appreciation, driven by subjective emotional responses and pleasure. Neuroaesthetic studies typically employ fMRI to measure brain activity during the observation of digital reproductions of paintings and sculptures (Pearce et al., 2016), while MEG (Chatterjee & Vartanian, 2014) and EEG (Di Dio & Vittorio, 2009; Zeki, 1999) have also been used to explore neuroelectric and neuromagnetic correlates of aesthetic perception. However, due to the immobility of fMRI and MEG, most studies present fine art images via screens rather than examining responses to real-world masterpieces. Modern EEG equipment enable field studies in real environments, such as art galleries (Babiloni et al., 2013). Di Dio et al. (2007) found that classical sculptures activated areas such as the lateral occipital gyrus, precuneus, and prefrontal cortex, with significant involvement of the right anterior insula when viewing original (versus altered) proportions. More recently, studies have leveraged portable EEG and biometric sensors to examine real-world aesthetic experiences. Babiloni et al. (2014) recorded EEG, heart rate (HR), and galvanic skin response (GSR) as participants viewed Michelangelo's Moses from different perspectives and lighting conditions, revealing that brain activity varied significantly based on vantage point, with emotional engagement peaking when Moses' gaze met the viewer's. Another study by Babiloni et al. (2015) used mobile EEG and eye tracking (ET) to analyze visitor responses to Tiziano Vecellio's paintings in a gallery setting, showing that portraits elicited higher approach-withdrawal (AW) indices than religious subjects, with eye fixation patterns correlating with aesthetic appreciation. Similarly, Cartocci et al. (2016) employed neurometric indices to assess responses to readings of Dante's Divina Commedia, finding that literary experts showed greater emotional and cognitive engagement than non-experts.

Although many of the studies presented in this section on aesthetics are related to the world of art, the results can be applied to the marketing aesthetics that we are dealing with in this book, since the brain activation mechanisms are unconscious reactions to beauty. With the aim of linking all the concepts presented above, the authors of this book have created a table with a list of brain areas of interest related to the characteristics of luxury aesthetics. Table 4.2 outlines the relationship between key brain regions and fundamental characteristics of luxury aesthetics. This growing body of research underscores the value of Neuroaesthetics in understanding consumer engagement with design, and luxury experiences.

Table 4.2 Brain regions and luxury aesthetics

Brain region	Relation to luxury aesthetic characteristics
Visual Cortex (Lateral Occipital Cortex)	Processes visual elements such as contrast, color, shapes, and clarity. Contributes to the perception of beauty and visual impact.
Orbitofrontal Cortex (OFC)	Involved in the emotional evaluation of beauty, particularly in judging stimuli as beautiful or ugly. Influences elegance, style, and overall aesthetic pleasure.
Prefrontal Cortex (Left Dorsolateral PFC)	Engages in higher-order processing and decision-making related to aesthetic judgment, impacting recognizability, craftsmanship, and design.
Frontal Cortex (Frontomedian Cortex)	Activated during the aesthetic judgment of beauty, contributing to the experience of elegance and cultural heritage.
Parietal Cortex (Posterior Parietal Cortex)	Involved in spatial awareness and the integration of visual information, aiding in the perception of harmony, unity, and order
Default Mode Network (DMN)	Supports the integration of emotional and cognitive processes, affecting the overall experience of beauty and originality.
Executive Control Network (ECN)	Facilitates top-down processing, influencing attention to detail, simplicity of lines, and minimalism in design.
Motor Cortex	Engages during the evaluation of disliked aesthetics, possibly linked to action-readied aversion, which contrasts with the appreciation of luxury.

Source: Authors' elaboration

Conclusion

This chapter has provided a comprehensive analysis of aesthetics in the luxury sector, integrating perspectives from marketing, the unique characteristics of luxury aesthetics, and the emerging field of Neuroaesthetics. The application of the neuroscience review in the field of aesthetics and luxury offers academics and managers an up-to-date view of the mechanisms that influence consumer behaviors, enabling luxury brands to create products and experiences that resonate deeply with consumers and increase their overall satisfaction and loyalty. Appealing emotionally in the design of products and services can be something that differentiates from other brands and drives preference. We have seen that the existing literature on aesthetics is more developed, but the luxury sector needs more specific studies, drawing on aesthetics research in other fields as a starting point.

Summary

The present chapter introduced the concept of aesthetics. Luxury aesthetics extend far beyond visual appeal as they engage deeply with the psychological and neurological processes that influence how individuals perceive and attribute value to objects. The sensory and emotional dimensions of luxury products create an enduring impact, stimulating cognitive and affective responses that transcend their physical attributes. This chapter has examined key aesthetic principles and dimensions—e.g., beauty, recognizability, craftsmanship, and cultural heritage—highlighting their fundamental role in shaping consumer perceptions. These dimensions are associated with activations in the human brain which highlights the importance of aesthetic design in the context of luxury.

Emerging findings in neuroscience reveal that some features such as linear designs evoke a sense of balance and order, engaging the prefrontal cortex and enhancing perceptions of timelessness and quality. In contrast, the strategic use of bold or novel design elements stimulates attention and fosters perceptions of uniqueness, appealing to the brain's reward and novelty detection systems. These insights demonstrate the link between aesthetics and the neurological mechanisms that underpin consumer preferences. Creating a model that integrates aesthetics into the marketing strategies of the luxury sector can be of key importance for luxury brands because consumer perceptions are shaped not only by the functional qualities of the product but also by other less tangible factors such as product design.

References

Amatulli, C., De Angelis, M., Costabile, M., & Guido, G. (2017). *Sustainable luxury brands: Evidence from research and implications for managers*. Springer.

Amatulli, C., De Angelis, M., & Donato, C. (2021). The atypicality of sustainable luxury products. *Psychology and Marketing, 38*(11), 1990–2005. https://doi.org/10.1002/mar.21559

Ariely, D., & Berns, G. S. (2010). Neuromarketing: The hope and hype of neuroimaging in business. *Nature Reviews Neuroscience, 11*(4), 284–292. https://doi.org/10.1038/nrn2795

Armstrong, K. M., Fitzgerald, J. K., & Moore, T. (2006). Changes in visual receptive fields with microstimulation of frontal cortex. *Neuron, 50*(5), 791–798.

Audrin, C., Brosch, T., Sander, D., & Chanal, J. (2018). More than meets the eye: The impact of materialism on information selection during luxury choices. *Frontiers in Behavioral Neuroscience, 12*(August), 1–12. https://doi.org/10.3389/fnbeh.2018.00172

Babiloni, F., Cherubino, P., Graziani, I., Trettel, A., Bagordo, G., Cundari, C., Borghini, G., Aricö, P., Maglione, A. G., & Vecchiato, G. (2014). The great beauty: A neuroaesthetic study by neuroelectric imaging during the observation of the real Michelangelo's Moses sculpture. 2014 36th Annual International Conference of the IEEE Engineering in Medicine and Biology Society, 6965–6968.

Babiloni, F., Cherubino, P., Graziani, I., Trettel, A., Infarinato, F., Picconi, D., Borghini, G., Maglione, A. G., Mattia, D., & Vecchiato, G. (2013). Neuroelectric brain imaging during a real visit of a fine arts gallery: A neuroaesthetic study of XVII century Dutch painters. 2013 35th Annual International Conference of the IEEE Engineering in Medicine and Biology Society (EMBC), 6179–6182.

Babiloni, F., Rossi, D., Cherubino, P., Trettel, A., Picconi, D., Maglione, A. G., Vecchiato, G., & Babiloni, F. (2015). A neuroaesthetic study of the cerebral perception and appreciation of paintings by titian using EEG and eyetracker measurements. Symbiotic Interaction: 4th International Workshop, Symbiotic 2015, Berlin, Germany, October 7–8, 2015, Proceedings 4, 21–32.

Baghirov, F., & Zhang, Y. (2024). Assessment of the association between aesthetic products and perceived product quality: An analysis of customer attitudes. *Journal of Consumer Marketing, 41*(7), 789–803.

Baisya, R. K., & Das, G. G. (2008). *Aesthetics in marketing*. SAGE Publications.

Bajaj, A., & Bond, S. D. (2018). Beyond beauty: Design symmetry and brand personality. *Journal of Consumer Psychology, 28*(1), 77–98. https://doi.org/10.1002/jcpy.1009

Balconi, M., Sebastiani, R., & Angioletti, L. (2019). A neuroscientific approach to explore consumers' intentions towards sustainability within the luxury fashion industry. *Sustainability (Switzerland), 11*(18). https://doi.org/10.3390/su11185105

Bellezza, S. (2023). Distance and alternative signals of status: A unifying framework. *Journal of Consumer Research, 50*(2), 322–342.

Berthon, P., Pitt, L., Parent, M., & Berthon, J.-P. (2009). Aesthetics and ephemerality: Observing and preserving the luxury brand. *California Management Review, 52*(1), 45–66.

Bigoin-Gagnan, A., & Lacoste-Badie, S. (2018). Symmetry influences packaging aesthetic evaluation and purchase intention. *International Journal of Retail & Distribution Management, 46*(11/12), 1026–1040. https://doi.org/10.1108/IJRDM-06-2017-0123

Bloch, P. H. (1995). Seeking the ideal form. *Journal of Marketing, 59*(Forty 1986), 16–29.

Bloch, P. H., Brunel, F. F., & Arnold, T. J. (2002). Individual differences in the centrality of visual product aesthetics: Concept and measurement. *Journal of Consumer Research, 29*(4), 551–565. https://doi.org/10.1086/346250

Boccia, M., Barbetti, S., Piccardi, L., Guariglia, C., Ferlazzo, F., Giannini, A. M., & Zaidel, D. W. (2016). Where does brain neural activation in aesthetic responses to visual art occur? Meta-analytic evidence from neuroimaging studies. *Neuroscience & Biobehavioral Reviews, 60*, 65–71.

Brasel, S. A., & Hagtvedt, H. (2016). Living brands: Consumer responses to animated brand logos. *Journal of the Academy of Marketing Science, 44*, 639–653.

Brodmann, K. (2007). *Brodmann's: Localisation in the cerebral cortex*. Springer Science & Business Media.

Buechel, E. C., & Townsend, C. (2018). Buying beauty for the long run: (Mis) predicting liking of product aesthetics. *Journal of Consumer Research, 45*(2), 275–297.

Bush, G., Luu, P., & Posner, M. I. (2000). Cognitive and emotional influences in anterior cingulate cortex. *Trends in Cognitive Sciences, 4*(6), 215–222.

Cartocci, G., Maglione, A. G., Modica, E., Rossi, D., Canettieri, P., Combi, M., Rea, R., Gatti, L., Perrotta, C. S., & Babiloni, F. (2016). The "NeuroDante Project": Neurometric measurements of participant's reaction to literary auditory stimuli from Dante's "Divina Commedia". *International Workshop on Symbiotic Interaction*, 52–64.

Carvalho, S. W., Hildebrand, D., & Sen, S. (2019). Dressed to impress: The effect of victim attire on helping behavior. *Journal of the Association for Consumer Research, 4*(4), 376–386.

Casado-Aranda, L. A., & Sanchez-Fernandez, J. (2022). Advances in neuroscience and marketing: Analyzing tool possibilities and research opportunities. *Spanish Journal of Marketing-ESIC, 26*(1), 3–22.

Cattaneo, Z., Lega, C., Ferrari, C., Vecchi, T., Cela-Conde, C. J., Silvanto, J., & Nadal, M. (2015). The role of the lateral occipital cortex in aesthetic appreciation of representational and abstract paintings: A TMS study. *Brain and Cognition, 95*, 44–53.

Cesareo, L., Townsend, C., & Pavlov, E. (2023). Hideous but worth it: Distinctive ugliness as a signal of luxury. *Journal of the Academy of Marketing Science, 51*(3), 636–657.

Charters, S. (2006). Aesthetic products and aesthetic consumption: A review. *Consumption, Markets and Culture, 9*(3), 235–255.

Chatterjee, A. (2011). Neuroaesthetics: A coming of age story. *Journal of Cognitive Neuroscience, 23*(1), 53–62.

Chatterjee, A., & Vartanian, O. (2014). Neuroaesthetics. *Trends in Cognitive Sciences, 18*(7), 370–375.

Chen, J., Wu, P., Jiang, Y., & Lv, L. (2025). Rough is durable, smooth is user-friendly: Effects of surface visual roughness on consumers' perception and product preference. *Journal of Retailing and Consumer Services, 84*, 104196.

Cherubino, P., Maglione, A. G., Graziani, I., Trettel, A., Vecchiato, G., & Babiloni, F. (2015). Measuring cognitive and emotional processes in retail: A neuroscience perspective. In *Successful technological integration for competitive advantage in retail settings* (pp. 76–92). IGI Global.

Cherubino, P., Martinez-Levy, A. C., Caratù, M., Cartocci, G., Di Flumeri, G., Modica, E., … & Trettel, A. (2019). Consumer behaviour through the eyes of neurophysiological measures: State-of-the-Art and future trends. *Computational Intelligence and Neuroscience, 2019*(1), 1976847.

Cho, J. Y., & Suh, J. (2020). Spatial color efficacy in perceived luxury and preference to stay: An eye-tracking study of retail interior environment. *Frontiers in Psychology, 11*(March), 1–15. https://doi.org/10.3389/fpsyg.2020.00296

Coates, D. (2003). Watches tell more than time: Product design, information, and the quest for Elegance. McGraw-Hill, New York.

Cohen, A. (2013). Kant on the possibility of ugliness. *British Journal of Aesthetics, 53*(2), 199–209.

Creusen, M. E. H., Veryzer, R. W., & Schoormans, J. P. L. (2010). Product value importance and consumer preference for visual complexity and symmetry. *European Journal of Marketing, 44*(9), 1437–1452. https://doi.org/10.1108/03090561011062916

Crolic, C., Zheng, Y., Hoegg, J., & Alba, J. W. (2019). The influence of product aesthetics on consumer inference making. *Journal of the Association for Consumer Research, 4*(4), 398–408.

Dai, X., Brendl, C. M., & Ariely, D. (2010). Wanting, liking, and preference construction. *Emotion, 10*(3), 324.

Dalgleish, T. (2004). The emotional brain. *Nature Reviews Neuroscience, 5*(7), 583–589.

Davidson, R. J. (2000). Affective style, psychopathology, and resilience: Brain mechanisms and plasticity. *American Psychologist, 55*(11), 1196.

Davidson, R. J. (2004). What does the prefrontal cortex "do" in affect: Perspectives on frontal EEG asymmetry research. *Biological Psychology, 67*(1–2), 219–234. https://doi.org/10.1016/J.BIOPSYCHO.2004.03.008

Davidson, R. J., & Irwin, W. (1999). The functional neuroanatomy of emotion and affective style. *Trends in Cognitive Sciences, 3*(1), 11–21.

Daw, N. D., O'doherty, J. P., Dayan, P., Seymour, B., & Dolan, R. J. (2006). Cortical substrates for exploratory decisions in humans. *Nature, 441*(7095), 876–879.

De Angelis, M., Amatulli, C., Pozharliev, R., Cardamone, E., & Di Cioccio, M. (2025). How Luxury Aesthetics Shapes Sustainable Consumption, working paper.

De Angelis, M., Amatulli, C., & Zaretti, M. (2020). The artification of luxury: How art can affect perceived durability and purchase intention of luxury products. *Sustainable Luxury and Craftsmanship*, 61–84.

Di Dio, C., Macaluso, E., & Rizzolatti, G. (2007). The golden beauty: Brain response to classical and renaissance sculptures. *PloS One, 2*(11), e1201.

Di Dio, C., & Vittorio, G. (2009). Neuroaesthetics: A review. *Current Opinion in Neurobiology, 19*(6), 682–687. https://doi.org/10.1016/j.conb.2009.09.001

Dion, D., & Arnould, E. (2011). Retail luxury strategy: Assembling Charisma through art and magic. *Journal of Retailing, 87*(4), 502–520. https://doi.org/10.1016/j.jretai.2011.09.001

Eichenbaum, H. (2000). A cortical–hippocampal system for declarative memory. *Nature reviews neuroscience, 1*(1), 41–50.

Floresco, S. B. (2015). The nucleus accumbens: An interface between cognition, emotion, and action. *Annual Review of Psychology, 66*(1), 25–52. https://doi.org/10.1146/annurev-psych-010213-115159

Godey, B., Lagier, J., & Pederzoli, D. (2009). A measurement scale of "aesthetic style" applied to luxury goods stores. *International Journal of Retail and Distribution Management, 37*(6), 527–537. https://doi.org/10.1108/09590550910956250

Grewal, L., Hmurovic, J., Lamberton, C., & Reczek, R. W. (2019). The self-perception connection: Why consumers devalue unattractive produce. *Journal of Marketing, 83*(1), 89–107.

Gupta, T., & Hagtvedt, H. (2021). Safe together, vulnerable apart: How interstitial space in text logos impacts brand attitudes in tight versus loose cultures. *Journal of Consumer Research, 48*(3), 474–491.

Hagtvedt, H. (2020). Art and aesthetics. In *Research handbook on luxury branding* (pp. 171–189). Edward Elgar Publishing.

Hagtvedt, H. (2022a). A brand (new) experience: Art, aesthetics, and sensory effects. *Journal of the Academy of Marketing Science, 50*(3), 425–428.

Hagtvedt, H. (2022b). Chapter 6 art and aesthetics in the future of luxury. In A. Joy (Ed.), *The future of luxury brands: Artification and sustainability* (pp. 115–134). De Gruyter. https://doi.org/10.1515/9783110732757-006

Hagtvedt, H. (2023). Aesthetics in marketing. *Foundations and Trends® in Marketing, 18*(2), 94–175.

Heitmann, M., Landwehr, J. R., Schreiner, T. F., & Van Heerde, H. J. (2020). Leveraging brand equity for effective visual product design. *Journal of Marketing Research, 57*(2), 257–277.

Hoyer, W. D., & Stokburger-Sauer, N. E. (2012). The role of aesthetic taste in consumer behavior. *Journal of the Academy of Marketing Science, 40*, 167–180.

Huang, M. H. (2001). The theory of emotions in marketing. *Journal of Business and Psychology, 16*, 239–247.

Janssen, C., Vanhamme, J., & Leblanc, S. (2017). Should luxury brands say it out loud? Brand conspicuousness and consumer perceptions of responsible luxury. *Journal of Business Research, 77*, 167–174.

Jiang, Y., Su, L., & Zhu, R. (2019). The shape of money: The impact of financial resources on product shape preference. *Journal of the Association for Consumer Research, 4*(4), 436–445.

Joy, A., Wang, J. J., Chan, T. S., Sherry, J. F., Jr., & Cui, G. (2014). M (Art) worlds: Consumer perceptions of how luxury brand stores become art institutions. *Journal of Retailing, 90*(3), 347–364.

Kapferer, J. N. (2014). The artification of luxury: From artisans to artists. *Business Horizons, 57*(3), 371–380. https://doi.org/10.1016/j.bushor.2013.12.007

Kapferer, J. N. (2017). Managing luxury brands. In J. N. Kapferer, J. Kernstock, T. Brexendorf, & S. Powell (Eds.), *Advances in luxury brand management. Journal of Brand Management: Advanced collections*. Palgrave Macmillan. https://doi.org/10.1007/978-3-319-51127-6_11

Kawabata, H., & Zeki, S. (2004). Neural correlates of beauty. *Journal of Neurophysiology, 91*(4), 1699–1705. https://doi.org/10.1152/jn.00696.2003

Keh, H. T., Wang, D., & Yan, L. (2021). Gimmicky or effective? The effects of imaginative displays on customers' purchase behavior. *Journal of Marketing, 85*(5), 109–127.

Kenning, P., & Plassmann, H. (2005). NeuroEconomics: An overview from an economic perspective. *Brain Research Bulletin, 67*(5), 343–354. https://doi.org/10.1016/j.brainresbull.2005.07.006

Kim, D., Hyun, H., & Park, J. (2020). The effect of interior color on customers' aesthetic perception, emotion, and behavior in the luxury service. *Journal of Retailing and Consumer Services, 57*(July), 102252. https://doi.org/10.1016/j.jretconser.2020.102252

Kim, H. S., Sherman, D. K., Ko, D., & Taylor, S. E. (2006). Pursuit of comfort and pursuit of harmony: Culture, relationships, and social support seeking. *Personality and Social Psychology Bulletin, 32*(12), 1595–1607.

Knudsen, E. I. (2007). Fundamental components of attention. *Annual Review of Neuroscience, 30*(1), 57–78.

Kumar, M. (2015). Aesthetic principles of product form and cognitive appraisals: Predicting emotional responses to beauty. In *The Psychology of Design* (pp. 234–251). Routledge.

Kumar, M., & Garg, N. (2010). Aesthetic principles and cognitive emotion appraisals: How much of the beauty lies in the eye of the beholder? *Journal of Consumer Psychology, 20*(4), 485–494.

Landwehr, J. R., McGill, A. L., & Herrmann, A. (2011). It's got the look: The effect of friendly and aggressive "facial" expressions on product liking and sales. *Journal of Marketing, 75*(3), 132–146.

Landwehr, J. R., Wentzel, D., & Herrmann, A. (2013). Product design for the long run: Consumer responses to typical and atypical designs at different stages of exposure. *Journal of Marketing, 77*(5), 92–107.

Lange, C. G. (1885). The mechanism of the emotions. *The Classical Psychologists*, 672–684.

Levy, S. J., & Czepiel, J. (1974). Marketing and aesthetics in Levy, S. J. (1999). *Brands, consumers, symbols and research: Sidney J Levy on marketing*. Sage Publications.

Li, D., Li, Y., & Wu, R. (2013). Consumer aesthetic experience in product design. *Advances in Psychological Science, 21*(2), 336–346.

Liu, Z. J., Yildirim, P., & Zhang, Z. J. (2024). Less is more: A theory of minimalist luxury. *Journal of Economics and Management Strategy, 33*(1), 78–110. https://doi.org/10.1111/jems.12558

Luchs, M. G., Swan, K. S., & Creusen, M. E. (2016). Perspective: A review of marketing research on product design with directions for future research. *Journal of Product Innovation Management, 33*(3), 320–341.

Luffarelli, J., Stamatogiannakis, A., & Yang, H. (2019). The visual asymmetry effect: An interplay of logo design and brand personality on brand equity. *Journal of Marketing Research, 56*(1), 89–103.

MacLean, P. D. (1985). Evolutionary psychiatry and the triune brain. *Psychological Medicine, 15*(2), 219–221.

Maren, S., & Quirk, G. J. (2004). Neuronal signalling of fear memory. *Nature Reviews Neuroscience, 5*(11), 844–852.

McGaugh, J. L. (2000). Memory – A century of consolidation. *Science, 287*(5451), 248–251. https://doi.org/10.1126/science.287.5451.248

Menninghaus, W., Wagner, V., Wassiliwizky, E., Schindler, I., Hanich, J., Jacobsen, T., & Koelsch, S. (2019). What are aesthetic emotions? *Psychological Review, 126*(2), 171.

Michael-Titus, A. T., & Shortland, P. (2022). *The nervous system, e-book: The nervous system*. E-Book. Elsevier Health Sciences.

Morris, R., Amaral, D. G., Bliss, T., Duff, K., & O'Keefe, J. (Eds.). (2024). *The hippocampus book*. Oxford University Press.

Nadal, M., & Vartanian, O. (2022). *The Oxford handbook of empirical aesthetics*. Oxford University Press.

O'Doherty, J. P. (2004). Reward representations and reward-related learning in the human brain: Insights from neuroimaging. *Current Opinion in Neurobiology, 14*(6), 769–776. https://doi.org/10.1016/j.conb.2004.10.016

Okonkwo, U. (2016). *Luxury fashion branding: Trends, tactics, techniques*. Springer.

Orth, U. R., & Wirtz, J. (2014). Consumer processing of interior service environments: The interplay among visual complexity, processing fluency, and attractiveness. *Journal of Service Research, 17*(3), 296–309.

Patrick, V. M., Peracchio, L. A., & Townsend, C. (2019). Introduction to special issue: Uncovering the potential of aesthetics and design to transform everyday life. *Journal of the Association for Consumer Research, 4*(4), 306–312.

Pearce, M. T., Zaidel, D. W., Vartanian, O., Skov, M., Leder, H., Chatterjee, A., & Nadal, M. (2016). Neuroaesthetics: The cognitive neuroscience of aesthetic experience. *Perspectives on Psychological Science, 11*(2), 265–279. https://doi.org/10.1177/1745691615621274

Pelowski, M., Markey, P. S., Forster, M., Gerger, G., & Leder, H. (2017). Move me, astonish me… delight my eyes and brain: The Vienna Integrated Model of top-down and bottom-up processes in Art Perception (VIMAP) and corresponding affective, evaluative, and neurophysiological correlates. *Physics of Life Reviews, 21*, 80–125.

Pombo, M., & Velasco, C. (2021). How aesthetic features convey the concept of brand premiumness. *Psychology & Marketing, 38*(9), 1475–1497. https://doi.org/10.1002/mar.21534

Pozharliev, R., Rossi, D., & De Angelis, M. (2021a). Anxious attachment style and consumer physiological emotional responses to human–robot service interactions. *Journal of Neuroscience, Psychology, and Economics, 14*(2), 59.

Pozharliev, R., De Angelis, M., Rossi, D., Romani, S., Verbeke, W., & Cherubino, P. (2021b). Attachment styles moderate customer responses to frontline service robots: Evidence from affective, attitudinal, and behavioral measures. *Psychology & Marketing, 38*(5), 881–895.

Pozharliev, R., Rossi, D., & De Angelis, M. (2022a). Consumers' self-reported and brain responses to advertising post on Instagram: The effect of number of followers and argument quality. *European Journal of Marketing, 56*(3), 922–948.

Pozharliev, R., De Angelis, M., & Rossi, D. (2022b). The effect of augmented reality versus traditional advertising: A comparison between neurophysiological and self-reported measures. *Marketing Letters, 33*(1), 113–128.

Pozharliev, R., Rossi, D., & De Angelis, M. (2022c). A picture says more than a thousand words: Using consumer neuroscience to study Instagram users' responses to influencer advertising. *Psychology & Marketing, 39*(7), 1336–1349.

Pozharliev, R., Verbeke, W. J., & Bagozzi, R. P. (2017). Social consumer neuroscience: Neurophysiological measures of advertising effectiveness in a social context. *Journal of Advertising, 46*(3), 351–362.

Pozharliev, R., Verbeke, W. J. M. I., Van Strien, J. W., & Bagozzi, R. P. (2015). Merely being with you increases my attention to luxury products: Using EEG to understand consumers' emotional experience with luxury branded products. *Journal of Marketing Research, 52*(4), 546–558. https://doi.org/10.1509/jmr.13.0560

Preuschoff, K., Quartz, S. R., & Bossaerts, P. (2008). Human insula activation reflects risk prediction errors as well as risk. *Journal of Neuroscience, 28*(11), 2745–2752.

Reimann, M., Schilke, O., Weber, B., Neuhaus, C., & Zaichkowsky, J. (2011). Functional magnetic resonance imaging in consumer research: A review and application. *Psychology & Marketing, 28*(6), 608–637.

Reimann, M., Zaichkowsky, J., Neuhaus, C., Bender, T., & Weber, B. (2010). Aesthetic package design: A behavioral, neural, and psychological investigation. *Journal of Consumer Psychology, 20*(4), 431–441.

Rilling, J. K., & Sanfey, A. G. (2011). The neuroscience of social decision-making. *Annual Review of Psychology, 62*, 23–48.

Roese, N. J., Melo, H., Vrantsidis, T., & Cunningham, W. A. (2017). Reward system. In *Consumer neuroscience* (pp. 207–219). MIT Press.

Romeo-Arroyo, E., Jensen, H., Hunneman, A., & Velasco, C. (2023). Assessing the influence of packaging design symmetry, curvature, and mark on the perception of brand premiumness. *International Journal of Gastronomy and Food Science, 31*, 100656.

Roux, E., Tafani, E., & Vigneron, F. (2017). Values associated with luxury brand consumption and the role of gender. *Journal of Business Research, 71*, 102–113. https://doi.org/10.1016/j.jbusres.2016.10.012

Salah, A. A. A., & Salah, A. A. (2008). Technoscience art: A bridge between neuroesthetics and art history? *Review of General Psychology, 12*(2), 147–158. https://doi.org/10.1037/1089-2680.12.2.147

Sanfey, A. G., Rilling, J. K., Aronson, J. A., Nystrom, L. E., & Cohen, J. D. (2003). The neural basis of economic decision-making in the ultimatum game. *Science, 300*(5626), 1755–1758.

Schnurr, B. (2017). The impact of atypical product design on consumer product and brand perception. *Journal of Brand Management, 24*(6), 609–621.

Schnurr, B. (2019). Too cute to be healthy: How cute packaging designs affect judgments of product tastiness and healthiness. *Journal of the Association for Consumer Research, 4*(4), 363–375.

Sevilla, J., & Townsend, C. (2016). The space-to-product ratio effect: How interstitial space influences product aesthetic appeal, store perceptions, and product preference. *Journal of Marketing Research, 53*(5), 665–681.

Shi, A., Huo, F., & Hou, G. (2021). Effects of design aesthetics on the perceived value of a product. *Frontiers in Psychology, 12*, 670800. https://doi.org/10.3389/fpsyg.2021.670800

Skov, M., Vartanian, O., Martindale, C., & Berleant, A. (2017). What is Neuroaesthetic? In Neuroaesthetics. https://www.taylorfrancis.com/books/9781351842860

Song, X., Jung, J., & Zhang, Y. (2021). Consumers' preference for user-designed versus designer-designed products: The moderating role of power distance belief. *Journal of Marketing Research, 58*(1), 163–181.

Su, L., Wan, E. W., & Jiang, Y. (2019). Filling an empty self: The impact of social exclusion on consumer preference for visual density. *Journal of Consumer Research, 46*(4), 808–824.

Sugam, J. A., Day, J. J., Wightman, R. M., & Carelli, R. M. (2012). Phasic nucleus accumbens dopamine encodes risk-based decision-making behavior. *Biological Psychiatry, 71*(3), 199–205. https://doi.org/10.1016/j.biopsych.2011.09.029

Townsend, C. (2017). The price of beauty: Differential effects of design elements with and without cost implications in nonprofit donor solicitations. *Journal of Consumer Research, 44*(4), 794–815.

Townsend, C., & Shu, S. B. (2010). When and how aesthetics influences financial decisions. *Journal of Consumer Psychology, 20*(4), 452–458.

Townsend, C., & Sood, S. (2012). Self-affirmation through the choice of highly aesthetic products. *Journal of Consumer Research, 39*(2), 615–628.

Treder, M. S. (2010). Behind the looking-glass: A review on human symmetry perception. *Symmetry, 2*(3), 1510–1543. https://doi.org/10.3390/sym2031510

Van Rompay, T. J. L., & Pruyn, A. T. H. (2011). When visual product features speak the same language: Effects of shape-typeface congruence on brand perception and price expectations. *Journal of Product Innovation Management, 28*(4), 599–610.

Venkatesh, A., Joy, A., Sherry Jr, J. F., & Deschenes, J. (2010). The aesthetics of luxury fashion, body and identify formation. *Journal of Consumer Psychology, 20*(4), 459–470.

Vigneron, F., & Johnson, L. W. (2017). Measuring Perceptions of Brand Luxury. In J. N. Kapferer, J. Kernstock, T. Brexendorf, & S. Powell (Eds.), *Advances in luxury brand management. Journal of Brand Management: Advanced Collections*. Palgrave Macmillan. https://doi.org/10.1007/978-3-319-51127-6_10

Wallis, J. D. (2007). Orbitofrontal cortex and its contribution to decision-making. *Annual Review of Neuroscience, 30*, 31–56.

Wang, R. (2024). A review of the impact of aesthetic experience on consumers' purchase decisions. *Highlights in Business, Economics and Management, 27*, 312–318.

Wang, R., Mo, F., Shen, Y., Song, Y., Cai, H., & Zhu, J. (2023). Functional connectivity gradients of the insula to different cerebral systems. *Human Brain Mapping, 44*(2), 790–800.

Wang, Y. (2022). A conceptual framework of contemporary luxury consumption. *International Journal of Research in Marketing, 39*(3), 788–803.

Wilson, A. V., & Bellezza, S. (2022). Consumer minimalism. *Journal of Consumer Research, 48*(5), 796–816.

Wilson, M. (2002). Six views of embodied cognition. *Psychonomic Bulletin & Review, 9*(4), 625–636.

Zanette, M. C., Pueschel, J. and Touzani, M. (2022). Re-arranging dressing practices: The role of objects in spreading ugly luxury. *Journal of Business Research, 145*, 784–800.

Zeki, S. (1998). ■ REVIEW: Parallel processing, asynchronous perception, and a distributed system of consciousness in vision. *The Neuroscientist, 4*(5), 365–372.

Zeki, S. (1999). Art and the brain. *Journal of Consciousness Studies, 6*(6–7), 76–96.

Zhang, W., Jin, J., Wang, A., Ma, Q., & Yu, H. (2019). Consumers' implicit motivation of purchasing luxury brands: An EEG study. *Psychology Research and Behavior Management, 12*, 913–929. https://doi.org/10.2147/PRBM.S215751

Zhao, L., Peng, J., & Yu, S. (2023). Sustainable luxury and consumer purchase intention: A systematic literature review. *SAGE Open, 13*(4), 21582440231216285.

5

Made in Italy Cases from Personal Luxury Goods and Services Industries

In Chap. 3, we supported the idea that luxury and sustainability are far more compatible than oxymoronic, and we focused our attention on both the drivers of sustainable consumer behavior as well as the sustainable initiatives of luxury brands. In Chap. 4, our attention was directed towards the fundamental principles of aesthetics and Neuroaesthetics within the luxury sector. While, across the previous two chapters we progressively support the idea that luxury organizations possess significant resources to embrace the challenge of sustainable development regardless of the sector in which they operate. However, knowing that luxury firms operate across different industries, we also underline that each of them features distinct environmental challenges and concerns. For instance, firms might focus on minimizing waste, improving worker conditions, or recycling more of their products, depending on their industry-specific standards and regulations. Similarly, each industry may have specific aesthetic standards, which require careful examination.

Interestingly, based on the most recent estimates provided by the Bain & Company-Altagamma Luxury Goods Worldwide Market Study, 80% of global luxury consumption in 2023 stemmed from three sectors:

A. D'Aniello et al., *Aesthetics and Sustainable Luxury Consumption*, Palgrave Advances in Luxury, https://doi.org/10.1007/978-3-031-96833-4_5

Table 5.1 Economic value of major luxury industries

Industry	Value (Billion)	% over total luxury consumption (2023)	Growth vs. 2022 (current exchange rates)	Growth vs. 2019 (current exchange rates)
Automotive	€635	42%	+12%	14%-16%
Personal luxury goods	€362	24%	+4%	28%-30%
Hospitality	€215	14%	+14%	3%-5%
Tot.	**€1.212**	**80%**		

Source: D'Arpizio et al. (2024)

automotive, personal luxury goods, and hospitality (Table 5.1; D'Arpizio et al., 2024).

Therefore, the present chapter specifically examines the luxury sectors that are considered highly significant due to their economic worth and projected environmental consequences.Specifically, we will differentiate among personal luxury goods (namely apparel, watches, jewelry, perfumes, and automotive) and luxury services which drive tourism flows (namely hotels, spas, and fine dining; e.g. Amatulli et al., 2021; Bellezza, 2023; Wang, 2022) in order to identify the specific aesthetic characteristics that characterize the development of every sector. Indeed, aesthetics is an essential component of luxury, as it facilitates the delivery of exceptional experiences (Kapferer, 2014; Venkatesh et al., 2010). Consequently, luxury brand managers invest significant resources to fulfil consumers' aspirations for pleasurable and visually captivating experiences (Amatulli et al., 2017). Luxury goods and service surroundings are exquisite, sophisticated, and visually appealing, rendering them nearly akin to a "work of art" (Venkatesh et al., 2010). Aesthetics profoundly impacts consumer choices (Townsend, 2017; Townsend & Shu, 2010) by eliciting emotional responses and enhancing the perceived worth, distinctiveness, and desirability of items or services (Holmqvist et al., 2020; Kapferer & Bastien, 2012). Throughout this chapter we will explore the specific challenges that each sector is facing while transitioning to more sustainable business models, presenting cases and evidence from the Italian market for both luxury goods and services.

Personal Luxury Goods and Automotive

The personal luxury goods sector is experiencing growth mainly because of continued price increases and relatively stable levels of consumption (D'Arpizio et al., 2024). According to the more recent estimates of Deloitte's annual study "Global Powers of Luxury Goods 2023" in 2022 the Top 100 luxury goods companies collectively made revenues of 347 billion dollars, which is an increase of 42 billion dollars more than 2021 (Deloitte, 2023). The list included 23 Italian companies (32.4% of the total), 19 of which are still owned by the founding family, confirming the crucial role of Italy as a leading geography for luxury productions. Hereafter we will review each major luxury good product category, focusing on relevant sustainable challenges as well as key aesthetic principles that drive the development of each sector.

Luxury Fashion

The production of fashion goods harms the environment in different ways: cotton needs a lot of water and non-organic cotton requires pesticides; fabric dyes can be loaded with toxins, and some luxury items require the use of endangered animals' fur or skins (Amatulli et al., 2017). Furthermore, the fashion industry is responsible for nearly 10% of greenhouse gas emissions and 20% of water waste, while consuming a significant amount of energy resources (i.e., more than aviation and shipping combined) (Cho, 2021). Indeed, the negative effects mostly stem from mass-market fast-fashion firms that heavily depend on mass production in nations with low-wage labor. In contrast, luxury fashion demonstrates a significantly higher level of consistency in its commitment to sustainability. This is primarily achieved through its emphasis on superior quality and durability, preference for local manufacturing, and preservation of creative craftsmanship and cultural heritage (Kapferer & Michaut-Denizeau, 2015; Kapferer & Michaut-Denizeau, 2019; Sun et al., 2021). In this sense, a fashion luxury brand can execute several sustainable activities without being inconsistent with traditional brand values. Moreover, the growing availability of luxury goods, known as the democratization of luxury, has led to significant effects in terms of costs, quantities, production time, and environmental sustainability (Karaosman et al., 2020). Consequently, the management of sustainability in luxury supply chains is becoming increasingly crucial due to the growing challenges faced by luxury enterprises in the mid-market fashion industry.

Through the analysis of the production processes of ten Italian luxury fashion brands from the textile and leather industries, Karaosman et al. (2020) identify some best practices for sustainability which they classify as being related to product, process, or supply chain activities.

Their study suggests that product level practices are focused on raw materials more than design activities, with often drive some operational benefits of cost reduction and market benefits, especially when firms are able to overcome potential barriers to sustainable sourcing (e.g., Bhandari et al., 2022). For example, the adoption of certified organic materials in the textile supply chain has become an increasingly popular trend that, despite higher initial material costs, yields long-term savings through reduced environmental impact and enhanced product quality. Given that quality is a direct determinant of market performance in luxury industries, clients' purchasing decisions are positively impacted, enhancing value: organic materials are now seen as advantageous due to their superior quality, which could be even reflected by pricing strategies, alongside cost savings from decreased water and chemical usage in processing (Karaosman et al., 2020). While leather companies were progressively involved in material evaluation and chemical limitations to guarantee product safety; yet, some conflicts arose between market demands for material's aesthetics, tactile qualities, and environmental concerns (Karaosman et al., 2020).

Process level measures were mainly focused on water management and energy efficiency, prompted by increasing apprehensions regarding pollution avoidance as well as cost reduction goals. Indeed, firms monitored their energy utilization, water consumption, and greenhouse gas emissions (Karaosman et al., 2020). However, integrating pollution protection strategies into corporate operations proved difficult. Controlling energy usage or carbon footprint in independent networks where the buying firm coordinates by outsourcing business services to third parties is complex.

Consequently, at supply chain level, luxury fashion brands are losing vertical integration, making supplier emission's control and energy monitoring harder. Because most organizations outsource their activities and raw material processing is hidden, supply chain complexity hinders the excellence of operational sustainability. For example, Italian silk production is primarily outsourced to Asia, thus it is not directly controllable. Therefore, in most firms, traceability initiatives and supplier audits were constrained by a lack of end-to-end supply chain visibility (Karaosman et al., 2020). Thus, improving supply chain traceability, often through suppliers' audit, is one key objective of sustainable luxury firms, since much of nature-based raw materials affect biodiversity destruction, increase water scarcity and present major challenges in terms of health and safety issues. However, product variety and suppliers' fragmentation represent a major challenge for a deep control.

Based on the above-mentioned analysis, the authors propose a framework of sustainability implementation (Fig. 5.1) within the luxury fashion industry, which is based on Italian's best practices and draws on stakeholder integration, innovativeness, and commitment as crucial antecedents of sustainable strategies at product, production, and supply chain level (Karaosman et al., 2020).

More generally, a fashion luxury brand can execute several sustainable activities without being inconsistent with traditional brand values. For example, the Italian brand Gucci, part of the Kering Group, launched the "Culture of Purpose" sustainability plan in 2015 to boost its long-term sustainability efforts in line with several environmental and social goals to be reached by 2025.[1] Drawing on the principles of the Triple Bottom

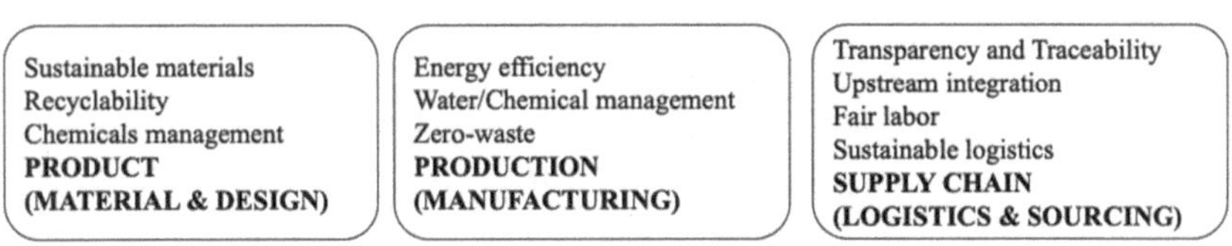

Fig. 5.1 Sustainability within the luxury fashion industry. (Source: Authors' elaboration)

[1] https://equilibrium.gucci.com/it/gucci-sustainability-strategy/

Line, Gucci's culture of purpose focuses on its goals for a sustainable development of People (e.g., promoting an inclusive culture, preserve artisanal traditions, support local communities, etc.) and Planet (e.g., reducing the environmental impact, lowering GhG emissions, increase traceability, etc.). Based on the framework of Karaosman et al. (2020) we briefly highlight some successful initiatives of this Italian brand. In terms of product level intervention, the brand announced it would abandon the use of animal furs starting in the Spring Summer Collection of 2018, committing to the international "Fur Free Retailer" program. In terms of process level measures, the brand launched the "Gucci-up" initiative for the creative upcycling of waste[2] and the "Gucci Scrap-less" project to minimize the impact of leather production through process innovation.[3] Finally, in terms of supply chain level intervention, the brand is able to ensure the complete traceability of 97% of its raw materials, while also making significant improvements to its use of sustainable materials such as recycled and organic textiles.[4] Meanwhile, Gucci has thoroughly invested in initiatives intended to support employees, local communities, and the broader society. For instance, the firm promotes the "École de l'Amour" to disseminate artisan skills and expertise to future generations and "Gucci CHIME" to boost gender equality.[5] Collectively, these initiatives position Gucci as a premier brand for sustainable development.

Certainly, fashion is the quintessential reflection of personal luxury. Renowned fashion houses such as Prada, Valentino and Dolce & Gabbana don't just sell clothes, they create sophisticated and exclusive lifestyles. Designers employ sumptuous materials like Italian silks, Scottish wools, and bespoke blends that elevate comfort and elegance (Amatulli et al., 2017). A fundamental attribute of an effective luxury fashion aesthetic design is its recognizability and distinctiveness. For example, Bottega Veneta, an Italian brand specialized in leather goods, is recognized for its unique production technique that allows workers to produce the distinctive weaving known as "intrecciato." This technique serves as a hallmark

[2] https://equilibrium.gucci.com/it/gucci-up/

[3] https://equilibrium.gucci.com/it/gucci-scrap-less/

[4] https://equilibrium.gucci.com/it/presenting-the-2022-gucci-equilibrium-impact-report/

[5] https://equilibrium.gucci.com/it/impactreport-2022/people/

of the brand, solidifying its connection to Italian tradition and superiority in leather craftsmanship (Amatulli et al., 2017).

In the field of culture and fashion theory, there has been a fair amount of attention paid to fashion and bodily appearance to recognize the aesthetics of production that includes dress, clothing, and costumes (Entwistle, 2023; Miller, 2007). Neither fashion nor aesthetics are timeless and therefore, the aesthetics of fashion is constantly evolving based on prevailing tastes and cultural dispositions (Slater, 2002). Indeed, the study by Venkatesh et al. (2010) reveals that luxury fashion brands significantly influence aesthetic values, body image, and identity formation. The research finds that these brands craft powerful visual narratives that shape perceptions of sophistication and status, impacting how individuals view their bodies and themselves. Luxury fashion often reinforces specific beauty standards and social hierarchies, which can affect personal self-expression and social positioning. The implications suggest that luxury brands should consider promoting more inclusive and diverse representations, while consumers should be aware of the impact fashion has on their self-perception and societal roles. Additionally, the study calls for a more culturally sensitive approach in fashion design and marketing, aiming to challenge rather than perpetuate existing stereotypes and exclusivities.

The essence of high-end fashion goes beyond the clothing and extends to the overall brand experience. Luxury boutiques, runway shows, and high-end packaging of exclusive design contribute to a positioning of desirability and sophistication, reinforcing the brand's elite status.

The concept of fashion aesthetics extends beyond just the clothing people wear to include other products like high-heeled shoes, which reflect artistic expression and empower the wearer. These adornments contribute to enhancing one's self-identity and social standing, providing individuals with a sense of prestige. In these scenarios, people view a combination of clothing and other accessories as desirable and fulfilling, thereby increasing the overall personal value and satisfaction for the consumer. Thus, fashion brands offer a central contribution to the customers' self-identity building process, mainly through their aesthetics, which is the result of the total looks such brands are able to share in the market.

Interestingly, given the importance of luxury fashion as a way to enhance individuals' self-identity, recent research, as well as popular press, focused on the so-called "ugly" luxury, that is about products that challenge the traditional standards of beauty (e.g., Cesareo et al., 2023; Sun et al., 2021; Zanette et al., 2022). While the latter can still be seen as a way to distinguish themselves from the mass, to signal their status (Bellezza, 2023), Zanette et al. (2022) suggest that research about consumers' embrace of ugly luxury should emphasize not only how these consumers attain distinction through possession, but also how they utilize the product and how the object influences their expressions of good taste.

Some emerging trends in luxury aesthetics include a shift towards more sustainable solutions such as recycled or upcycled materials (Adıgüzel & Donato, 2021). One prominent example is the Prada Re-nylon[6] collection, which results from a collaboration of the Italian luxury brand with Aquafil, an Italian textile yarn producer with over half a century of expertise in creating synthetic fibers. Indeed, the Re-nylon material, called "ECONYL®" is made of fishing nets, discarded nylon, carpet, and industrial waste, thus on the direct collection of these materials from landfills and oceans across the world. At the same time, however, luxury firms should be aware of the potential damage of integrating sustainable design, since different consumers may perceive eco-friendly products as aesthetically unappealing (Han et al., 2017). Consequently, luxury firms should focus on highlighting the coherence with their essential characteristics that include accuracy, artisanal components, and timeless design.

In this perspective, one example of sustainable luxury and aesthetics is Brunello Cucinelli, which made the concept of "beauty" a prominent foundation of its business processes. The case of this major Italian brand is described in detail in the following box: **The "Gentle luxury" of Brunello Cucinelli**.

---

[6] https://www.pradagroup.com/en/sustainability/environment-csr/prada-re-nylon.html

The "Gentle luxury" of Brunello Cucinelli

The expression used by Brunello Cucinelli, "Gentle luxury, a symbol of tailored beauty" fully captures the brand's philosophy regarding the value-driven meaning of luxury. Within this idea, we see how luxury is intrinsically linked to the values of kindness, beauty, and moderation, which are also central to the company's philosophy, as Chairman Brunello Cucinelli himself explains: "The idea of gentle luxury was born in me as the most faithful expression of our idea, of our creations, of our style. I see luxury as a symbol: it appears symbolic to me because it is something that binds us to beauty. Gentle luxury, more simply, is a friend who welcomes you with a sincere smile, it is someone who reaches out to you, it is creativity and genius that do not cross the line, it means finding beauty in simplicity. It is welcoming to all, it is within measure and is always appreciated, as it knows the value of respecting others. This is why gentle luxury has a universal value".

*Thus, the concept of "**beauty**" becomes central to the Casa di Moda's actions, serving as an inspiring focus for a series of concrete initiatives that encompass workspaces and beyond. These include the restoration of the medieval hamlet of Solomeo, which began over 30 years ago with the Castle (the company's historic headquarters), and extend to the creation of new spaces dedicated to promoting Art and Culture, such as the Cucinelli Theater, the Aurelian Neo-Humanistic Academy, The School of Contemporary High Craftsmanship and Arts, and the forthcoming Universal Library of Solomeo. Another crucial aspect is the landscape project, "Project for Beauty," which seeks to recover and enhance the green areas surrounding the hamlet; this project spans around 100 hectares divided into three areas: the Industrial Park, where the company headquarters are located, surrounded by a vast garden; the Don Alberto Seri Park, with green areas and a small stadium for the community's recreational activities; and, finally, the Agricultural Park, with a vineyard, winery, olive grove, oil mill, and the Monument to Human Dignity.*

The current company headquarters was developed with the goal of revitalizing an area previously occupied by seven industrial factories, covering around seven hectares. The architectural project, known as the "Beautiful Factory", is composed of spacious areas surrounded by large windows overlooking green spaces and fountains, with the goal of creating welcoming spaces within the company (including the pavilions dedicated to production, warehousing, and logistics) where collaborators can work in a relaxed, airy, and bright environment.

There is a particular episode in the youth of Chairman Brunello Cucinelli that marked a turning point in his life and, consequently, in that of the company as well: after spending his childhood in the countryside, at around fifteen, he moved to the city with his family. At that time, his family had lived in a small town in the Province of Perugia, near Lake Trasimeno, working the land. Brunello Cucinelli remembers these early years as a very happy time in his life, where, despite the lack of electricity at home, the family was united, there was solidarity with other local farmers, and everyone lived by the rhythm of the seasons, in harmony with nature. However, in search of better living conditions, his father found work in a factory, leading the whole family to move to one of the suburbs of Perugia. One day, his father came home from work distressed, with teary eyes, having been humiliated at work. This image profoundly struck Brunello, who from that moment on said to himself, "I don't yet know what I'll do in life, but I will try to pursue it for the moral and economic dignity of human beings". This awareness fostered the values of Humanistic Capitalism and Human Sustainability, which still serve as the ethical pillars of the company today.

Indeed, these values are reflected in every aspect of the company's life, primarily in the products and their life cycle. The importance of artisanal tradition and the exceptional level of professionalism of Italian artisans represent a global benchmark for manufacturing and design beyond just the fashion sector. Recognizing the significance of such a historical, cultural, industrial, and creative heritage, the Casa di Moda has chosen, from the very beginning, to establish every stage of its production in Italy, fostering relationships of respect and trust with highly skilled artisans who, today as in the past, experiment and innovate, always grounded in traditional techniques and the rich manufacturing heritage of Italy. In this light, the value of Italian craftsmanship has always been a key pillar of Brunello Cucinelli's brand. The production, entirely based in Italy, has allowed the Casa di Moda to create over time a high-level collaborative network with artisan workshops, each specializing in distinct and specific stages of the production process. This approach has enabled to build productive relationships with centers of excellence located in Umbria and other Italian districts, ensuring top-tier quality throughout all stages of the production chain.

The approach to product design adheres to some key principles that can be found in each collection: the combination of the highest quality raw materials, specialized craftsmanship, and distinctive design gives rise to products that transcend seasons and momentary trends, achieving a dimension of "timeless" elegance. This timelessness embraces the brand's unique stylistic codes, continuously evolving while remaining always true to itself. Themes that unite aesthetics with sustainability and responsible consumption are reflected in the "longevity by design" principle, an essential guideline for Brunello Cucinelli's garments. Beginning with the selection of raw materials, each piece is created to blend quality, functionality, style, and durability, offering customers a product to wear today and, ideally, to pass down from generation to generation. In this spirit, the company is pleased to offer a complimentary repair service, available to customers for any purchased product, with no time restrictions. The CARE[7] *service thus becomes a central theme, connecting the repair and durability of products to the company's broader strategic vision. Caring for creations is also a gesture of gratitude toward the artisans who crafted them.*

Source: The Authors thank Brunello Cucinelli S.p.A. for the interview.

Jewelry and Watches

Luxury jewelry and watches are bought and worn because of their durability and investment potential. According to the 2023 Jewelry Consumer Trend Report, which shapes jewelry consumption in China, about 60% of consumers consider the value preservation of gold jewelry when purchasing, and as the younger generation becomes the new main force of consumption, the demand for diversity and customization of jewelry consumption is increasing more and more (Chow Tai Fook and Ipsos, 2023). This is consistent with the notions of conscious and sustainable consumption discussed above. However, the production of these items heavily depends on the extraction of metals and gems, which has been historically linked to mining region degradation, poor working conditions, and water contamination (De Angelis et al., 2021). To address

[7] https://shop.brunellocucinelli.com/it-it/care.html

those concerns, several brands in this sector have adopted and communicated sustainable policies. For example, Tiffany & Co (tiffany.com/sustainability) aims to achieve 100% traceability of individually registered diamonds and precious metals, ensure sustainability in its material sourcing, adopt recycled materials, and further integrate sustainability into its design and innovation processes. At the same time, the brand aims to become the most inclusive luxury brand, launching initiatives related to employee diversity, wellbeing, and skill development. Finally, the brand aims to achieve net-zero greenhouse gas emissions, build sustainable stores, and completely eliminate single-use plastic for packaging and product conservation.

Exclusivity for luxury jewelry and watches can be manifested by physical stimuli, like the design of brand-owned stores in strategically selective sites, interaction with trained staff, or through the refined aesthetics of the product (Dinh et al., 2024). Indeed, both jewelry and watches are core elements of individual adornment, often considered like family treasures that are passed down through generations (Amatulli et al., 2017). The aesthetic appeal lies not only in the quality of the pieces, but also in the unique designs that highlight the distinctiveness of the consumer who wears them.

Luxury watches probably represent one of the main product categories that allow consumers to communicate the social group they belong to (Rovai, 2016). Indeed, luxury watches are elitist and exclusive products resulting from a deep social and historical understanding of luxury (Sukharevska, 2016). For instance, Italian brands such as Panerai and Bulgari produce watches that stand out for their movements and innovative technology. The longevity and eternalness of luxury watch models are characteristics of the high quality (Chattalas & Shukla, 2015). The watch model *Datejust* by Rolex may serve as a specific example. The design of this luxury watch model has nearly not changed since 1945 (Kapferer, 2017). The aesthetic of a luxury watch leads by means of its precision, functions, and design to an experience value which appeals to all senses (Kapferer, 2017). Because of its classiness, its product, and packaging design, its first-class interior at the point of sales and the myth which surrounds the brand of the specific luxury watch, it creates an aesthetic which ordinary watch brands cannot achieve (Kapferer, 2017). Godart et al. (2020) explored how aesthetics and technological complexity affect the evaluation of luxury watches among different consumers (namely professionals vs. others), defining aesthetic complexity as the presence of various design elements (e.g., material, case, band style, etc.). Their study revealed that professionals demonstrate higher aesthetic sophistication, while laypersons prefer moderate complexity (Godart et al., 2020). Luxury watch brands possess a history and therewith a brand heritage, associated with values that last for a long period of time (Munz, 2018, p. 198). So, luxury watches, with their rich history, precision engineering, and unparalleled aesthetic, stand as symbols of craftsmanship and tradition. The emotional connection they forge with consumers is rooted in their legacy and the values they represent. However, in a rapidly evolving market, the rise of technological innovations like smartwatches presents a new challenge: the future of watchmaking will likely need to balance timeless tradition with the innovation demanded by the digital age. In this context, it is essential to create a valuable consumer experience when purchasing products such as jewelry and watches. Adopting innovative approaches and crafting in physical stores and online experiences can improve brand visibility and consumer engagement, ultimately driving growth and competitive advantage in the luxury jewelry sector.

Cosmetics and Perfumes

The use of toxic ingredients and the animal testing of cosmetics and perfumes are one of the key sustainable concerns of consumers (Amatulli et al., 2017). Organic products are becoming more popular since consumers are more attentive to materials that came in contact with their bodies (Ruokamo et al., 2022). Similarly, inclusive products in terms of appearance (i.e., skin color, hair) or disabilities (e.g., including Braille instructions on packaging; Kolling et al., 2022), within the cosmetics industry, are of great importance nowadays. A brand example for this is "Fenty Beauty", created by the popstar Rihanna in 2017 to meet the demand for new make-up products adequate for all skin types and tones. The brand also promotes sustainability principles through packaging reduction and the use of "clean" (i.e., vegan, gluten-free) ingredients.

Shiseido is a Japanese brand that earned a reputation for economic, environmental, and social responsibility, driven by its mission to create "beauty innovations for a better world": from 2021 to 2023 the brand has been included in the Dow Jones Sustainability World Index, receiving optimal evaluations for governmental, environmental, and societal policies.

Within the luxury sector it is important not only to focus on creating products for beauty, but also to create a sensory pleasure experience. Some brands as Acqua di Parma, Gucci, and Armani introduce products that promise sensory delight. To do so, the packaging and its aesthetic appeal are crucial. The products themselves are developed using scientific advancements and quality ingredients, ensuring that their use provides long-lasting results. Textures and fragrances are carefully chosen to make an experience that is both luxurious and useful for all the senses. Over the years, the beauty and cosmetics industry has become more appealing to businesses from other fields because of its great growth potentialities and the relatively simple penetrability of its markets is relatively simple to penetrate (Leung et al., 2019). On this behalf, several luxury fashion brands have taken advantage of this opportunity by offering great brand extensions into such sector (Som & Blanckaert, 2015; Stankeviciute & Hoffmann, 2011) venturing into a new sector while preserving their identity, driving financial growth, and maintaining exclusivity (Kapferer, 2012). Entering a new market in this manner, referred to as category extension, is typically less common than line extensions (Dall'Olmo Riley et al., 2015). However, it is particularly popular among luxury brands such as Yves Saint Laurent, Dior, Chanel, and Lancôme to introduce cosmetics lines, as it is a reliable strategy that aligns well with their original brand image (Stankeviciute & Hoffmann, 2011). Perfumes from luxury brands transform the act of scent application into a form of art. Interestingly, recent research (Krishnan et al., 2022) sought to identify the factors that influenced consumers' intentions to purchase luxury-brand perfumes. By examining how elements such as uniqueness, brand reputation, cost, quality, and status affected consumer buying intentions for luxury fragrances, this study aimed to aid in the development of effective marketing strategies. The results revealed that consumers associated high-priced premium fragrances with a sense of prestige.

Moreover, the research on luxury fragrance brands in the Spanish market, conducted by Díaz-Bustamante et al. (2015), highlights several key attributes that contribute to the perception of luxury in this sector. These attributes include high prices, luxurious packaging, association with prestigious designers or creators, belonging to an already luxurious brand, exclusivity, and the fragrance's ability to seduce. Each of these elements directly influences values such as usage, uniqueness, quality, identity reaffirmation, materialism, hedonism, and social prestige. Higher prices and exclusive packaging enhance the perceived quality and uniqueness, while a link to renowned designers boosts social prestige and personal identity. The overall perception of luxury is heightened when fewer people use the fragrance, aligning it with exclusivity and prestige. This study is significant in understanding the aesthetics of luxury fragrances, as it provides a comprehensive model to analyze and improve the market positioning of these brands, ensuring they better align with the ideal brand attributes perceived by consumers.

Automotive

The automotive industry has proven to be one of the most harmful sectors for the planet due to the huge environmental impact of producing and using personal vehicles. The ownership of cars within cityscapes during the last century is part of the experience that individuals have throughout their daily routine (Hoffmann & Akbar, 2023; Pojani et al., 2018). Fittingly, then, this industry has been among the most innovative in terms of sustainable initiatives: encompassing carbon emissions, as well as the use of raw materials and chemicals throughout the production process. Since 2003, the electric car manufacturer Tesla has tried to do for the luxury market what Toyota did for mass-market cars in 1997: namely, popularize alternative fuel sources (Amatulli et al., 2017). Beyond their burgeoning interest in hybrid and fully electric vehicles, luxury customers are also looking for ultra-customized solutions (D'Arpizio et al., 2024), and luxury car manufacturers are focusing on innovation and exclusivity in interior design. Traditional materials (e.g., leather) are being replaced by more sustainable materials such as recycled plastic bottles, recycled nylon waste from fishing nets, sustainably sourced wood trim

(Williams, 2023), or vegan interiors (e.g., BMW's innovative cactus fibers; bmwgroup.com/sustainability). The Italian luxury brand Ferrari designed a sustainability strategy that involves five pillars: (1) sustainable technological innovation and excellence, (2) good governance driven by ethical considerations, (3) employees' wellbeing, through inclusion, education, and dedicated initiatives, (4) reduction of environmental footprint, and (5) boosting positive social impact.[8] For example, the firm launched "Bosco Ferrari", a forestation project in the Province of Modena, close to Maranello, which seeks to restore vegetation in a 30-hectare area and provide social value by employing local young people to take care of the newly planted trees.[9]

Aesthetics has consistently been a crucial part of design in the realm of premium automobiles. The appearance of the vehicle defines the identity and the emotional connection with its owner. Luxury cars have capitalized on meticulously designed visual aspects to convey strength, beauty, and exclusivity for decades. For example, the exterior design of cars is extremely relevant: sleek, aerodynamic lines traditionally convey speed and modernity, while more robust, angular shapes suggest strength. Even if design choices are crucial, nowadays the importance of sustainability influences those choices. Indeed, many luxury car markers have incorporated aerodynamic shapes for aesthetics and the ability to improve energy efficiency. This change demonstrates how aesthetics and sustainability can go hand in hand, with design elements such as the silhouette of a car becoming both a visual and functional statement.

Inside the car, aesthetics is also crucial for consumer preferences; specific materials, or some combinations of them, may create a special sense of comfort and luxury. With this regard, with the rise of sustainability, brands are choosing materials that, at the same time, maintain a luxurious feel and reduce environmental impact of production.

The color and the sound of cars are also part of their aesthetics and can be differently perceived by consumers in the luxury sector. For example, top-quality sports cars have always been loud, while hybrid and electric

---

[8] https://www.ferrari.com/en-EN/corporate/sustainability-strategy

[9] https://www.ferrari.com/en-EN/corporate/articles/bosco-ferrari-expands-with-new-forestation-initiative-in-formigine-corp

models stand out for their quietness. In terms of material innovation, Cervellon and Shammas (2013) report that Porsche cars are built from recyclable materials and in production facilities in which 100% of waste is recycled. Just as beauty and performance remain fundamental to design, there is a growing recognition that such values must now coexist with environmental responsibility. This evolution does not mean that the visual appeal of luxury cars will diminish; rather, it opens new possibilities for innovation, where style, performance, and sustainability are combined to create vehicles that are as responsible as they are desirable.

Luxury Services

While personal luxury goods and cars continue to dominate the global luxury market, the richest customers are progressively shifting their focus from material possessions to experiential purchases (Eckhardt & Bardhi, 2020). Indeed, in our experience-driven culture, luxury services satisfy the demand for immersive and unforgettable experiences, playing a significant role in today's economy and society. Exactly like products, they serve as symbols of prestige, quality, and exclusivity, being a major contributor to global economic growth, generating significant revenue and employment opportunities. Luxury travel, for example, drives significant tourism revenue and supports local economies by creating jobs in hospitality, dining, and leisure sectors. Fine dining services, exclusive and exotic vacations, luxury spas, and luxury hotels (e.g., Amatulli et al., 2021; Bellezza, 2023; Wang, 2022) have recently recovered from the Covid-19 crisis, returning to historical peaks, driven by a revival in social engagements and travel (D'Arpizio et al., 2024).

Hereafter we will review two major luxury service domains, namely hotels and fine dining, focusing on relevant sustainable challenges and drivers of aesthetic experiences. Indeed, the concept of servicescape, which refers to the physical surroundings where services are provided, entails the essence of beauty in service environment design (Bitner, 1992). The latter, which is undoubtedly related to visual aesthetics, is core in shaping consumer responses especially in terms of affective states of arousal and pleasure (Kumar et al., 2017) for different types of service settings.

Hotels and Spas

While the influx of tourists might bring several economic benefits to a geographic area, it also raises problems over energy consumption, excessive water usage, and trash management (e.g., Amatulli et al., 2021; Sourvinou & Filimonau, 2018).

Although those issues abated during the pandemic due to lockdowns causing a drastic decline in tourism, consumers' interest in unique and personalized travel experiences is resurging. However, many consumers no longer just want an escape from everyday life; they are also looking for authentic and culturally immersive experiences that respect both human and environmental health (Amatulli et al., 2021; D'Arpizio et al., 2024; Deloitte, 2023). Fortunately, hotels' varied operations and scale allow them to tackle sustainability from multiple angles (Amatulli et al., 2017). A good example is the new Bulgari Hotel located in Rome, which has heavily integrated sustainability and social responsibility into its design, dining services, and social inclusion activities. The hotel maintains more than four thousand plants all around the building, creating the impression of an urban jungle. Furthermore, it has donated over a million euros to restore the Mausoleum of Augustus and the Ara Pacis Museum, which are located close to the hotel. The hotel also utilizes renewable energy sources and local ingredients, preserves local art, and supports local craftsmen and workers.[10]

Pijls et al. (2017) explored the meaning of hospitality from consumers' perspective by taking the whole servicescape into account, including the physical dimensions of the service environment. Their research resulted in the Experience of Hospitality Scale, consisting of three factors: inviting, care, and comfort. Inviting refers to the experience of inviting, openness and freedom, care is about experiencing empathy, servitude, and acknowledgement. Comfort refers to feeling at ease, relaxed, and comfortable. In this context, aesthetics plays a crucial role in shaping consumers' responses towards sustainable services spaces. Literature on the effects of design characteristics of buildings on people's experience of hospitality suggests that aesthetics is a key factor in attracting people to

[10] https://www.bulgarihotels.com/en_US/rome/the-hotel/green-initiatives

buildings. For example, through a virtual reality experiment Pijls et al. (2023) found that transparent entrances significantly increase perceptions of hospitality across the three dimensions of the EH-scale (i.e., inviting, care and comfort; Pijls et al., 2017), since they were perceived as more beautiful, modern and attractive. When designing service environments, attention to aesthetics is particularly important (Kirillova & Chan, 2018). Designers may focus on visual design, with the main objective of increasing attractiveness (Grewal et al., 2003; Kirillova & Chan, 2018; Orth & Wirtz, 2014), especially when applied to the exterior design of the building (Baker et al., 2020; Lecointre-Erickson et al., 2024). Kim and Moon (2009) show that an attractive environment will lead to more positive emotions than an unattractive environment. Orth and Wirtz (2014) state that an appealing service environment captures attention, triggers approach behaviors, strengthens consumer attachment, and enhances the customer experience.

West & Hughes (1991, p. 364) stated "*design in hotels is concerned with the construction of tangible attributes that can be shown to and seen by potential customers: thus, relaying information to them about how the hotel views itself*". As we are dealing with service settings, environmental atmosphere and interior design are crucial to create positive experiences (Alfakhri et al., 2018). For example, renowned Italian luxury hotel such as Baglioni Hotels are celebrated for their elegance and contemporary Italian style. Consumer perception of luxury is strongly influenced by these aesthetic elements, as they associate the physical environment with feelings of well-being, status, and aesthetic pleasure. This perception is further enhanced when a hotel's design aligns with expectations of personalization and authenticity, creating an atmosphere that reinforces brand identity and elevates the guest experience to exceptional levels of satisfaction. In this sense, design can create a sense of place and emotional connection, setting a benchmark for luxury hospitality worldwide.

Customers select hotels based on cognitive (e.g. price, food, and service quality), sensory (e.g. general atmosphere, room quality), and affective (e.g. comfort, safety, and entertainment) attributes. In the hotel context, it has been revealed that consumer's sense making can be supported by aesthetics related to the physical environment (Norman, 2004; Rafaeli & Vilnai-Yavetz, 2004). Meanwhile, fashionable design is particularly important to a growing breed of discerning, style-conscious travelers (Bovich, 2007; Lim & Endean, 2009). Kim et al. (2020) studied the impact of interior color on aesthetic perceptions and suggest that luxury brand managers could leverage different colors, based on the values that they aim to communicate: dark brown for heritage, white for cleanliness, gold/silver for affluence. Project simplicity and elegance as well as a home feeling with comfort are also often able to entice customers (Chan et al., 2021). In addition, contemporary luxury design together with preservation of the property's traditions and heritage are the two key aesthetic elements that customers look for in boutique hotels (Lim & Endean, 2009).

Hereafter we describe the unique case of Borgo Egnazia, an esteemed symbol of Italian excellence in hospitality both in terms of luxuriousness and sustainable design.

Borgo Egnazia: Nowhere else

Borgo Egnazia is located in the Itria Valley of Apulia, between the Adriatic Sea and vast, historic olive fields. Since 2010, Borgo Egnazia has provided a distinctive hospitality concept rooted in the region's ancient culture, alongside a progressive and modern interpretation of tradition. Therefore, it serves as a universal emblem of Apulian hospitality, significantly enhancing the region development as a premium tourism destination.

The core values of Borgo Egnazia have always been rooted around a vision of luxury that transcends the externalization of consumption to show off status and wealth, emphasizing instead a profound commitment to each guest, providing tailor-made, exclusive, and distinctive experiences. Their motto "*Nowhere Else*" reflects this precisely: a special place where the most precious asset lies in people and the meticulous attention to each detail of the stay. All employees, including local consultants, waitstaff, and artisans, contribute to crafting personalized experiences that immerse guests in a genuine and welcoming atmosphere.

This philosophy also carries a strong sustainable element: Borgo Egnazia's approach to hospitality does not rely on excess or showiness, but instead embodies a more intimate and modest approach, rooted in people and its local land.

The strong bond between Borgo Egnazia and its surrounding area is woven into every aspect of the guest experience, encouraging visitors to step beyond the hotel and explore the beauty of the region. This connection is brought to life through a curated range of activities that bring guests in the heart of Apulia's landscapes, culture, and heritage. For instance, a bicycle excursion through the timeless groves of ancient olive trees, where guests can see the Apulian countryside and visit traditional underground oil mills, where the stones still carry the marks of centuries-old olive oil production. Here, guests can taste the "golden yellow" of freshly pressed oil.

Furthermore, the aesthetic design of Borgo Egnazia spaces embodies its profound connection to the region, exemplified by the use of local materials such as tuff, a simple and authentic stone with energy-efficient characteristics, that integrates well into the surrounding landscape. Likewise, wood is extensively utilized to evoke the heritage and customs of Apulian ancestors, while giving spaces a warm and natural atmosphere. The aesthetic design of the service spaces is crucial to the brand identity, as it deeply reflects the concept of Made in Italy, renowned for its "beautiful and well-made" productions. Borgo Egnazia was designed to respect the traditions, with colors harmonizing with the landscape and the history of nearby villages and traditional farmhouses typical of the Itria Valley. Shades of white and ochre dominate its spaces, telling tales of a rich and fascinating past of a traditional "masseria". Guests appreciate not only the beauty of its spaces but also the commitment to reflect the culture and aesthetics of the region. Similarly, products used at the SPA also exemplify this philosophy: the majority of treatments are crafted using natural, locally sourced ingredients, designated in Apulian dialect to reflect ancient methods and traditions.

Over time Borgo Egnazia has built a distinctive brand, known for its uniqueness and for offering a different experience compared to traditional 5-star luxury hotels. Clients go there to seek a discreet, authentic, luxury experience, which is far removed from the values of opulence and extravagance. These elements of uniqueness allow the company to attract a clientele that naturally values its vision of sustainability: Guests are generally in search of high-quality experiences and share respect for the land, local traditions, and the environment.

In this sense, luxury and sustainability are thoughtfully in communication with one another, within an exclusive setting that moves away from conventional norms of ostentatious consumption.

Borgo Egnazia's hospitality concept is rooted in a respectful engagement with history, culture, and the local environment—and the country-of-origin effect associated with "Made in Italy" further strengthens this favorable connection. Indeed, Made in Italy is widely acknowledged as a hallmark of quality and authenticity, particularly in Southern Italy, where craftsmanship, cuisine, and design are not merely economic activities, but rather represent forms of cultural expressions and art. In Apulia, luxury is intimately connected to respect for the land. It is a concept that values excellence, uniqueness, and a strong bond with the territory: From craftsmanship passed down through generations, to cuisine that uses local, seasonal ingredients, to design that incorporates traditional materials like tuff and wood, everything is intrinsically tied to nature and the reverence for the land's offerings.

To add even more value to the importance of cultural heritage, Borgo Egnazia recently founded an association called Clara,[11] dedicated to protecting and promoting the artisanal techniques of the local area. At Borgo Egnazia, these values manifest in utmost respect for the environment and resources, and a strong sense of give-back which is rooted in the company's culture. This approach not only contributes to making its concept of hospitality "sustainable and responsible" but also makes the experience more authentic, offering guests the true essence of Made in Italy, combined with a modern and sustainable vision.

Borgo Egnazia was a pioneer in revitalizing and reinterpreting the "old" architecture of the typical Apulian "masseria" into an innovative hospitality concept, serving as a trendsetter for other suppliers in the region. Indeed, many traditional farmhouses located in the Itria Valley have now become 5-star hotels, maintaining that natural and authentic sense of place, without any disruptive transformation of the land. In other words, Borgo Egnazia vision has fostered a balance between tradition and innovation, offering an experience that celebrates the local

[11] https://associazioneclara.org/

cultural heritage while embracing responsible and sustainable tourism in tune with the surrounding environment.

One of the greatest challenges that the brand faces is to continuously retain the territory, its culture, and its highest aesthetic expression. Initiatives like the DMC Indigenus,[12] which promotes authentic experiences, is an example of how the brand strives to keep local culture alive. Additionally, Local Advisers provide guests with personalized guidance during their stay. These aspects not only contribute to creating a unique and memorable atmosphere but are also elements that guests perceive and embrace with great joy.

The way luxury is defined at Borgo Egnazia is primarily based on the "care" for each guest, reflecting a deeper sense of hospitality. Details and meaningful experiences leave no space for excess. This attention to service and aesthetics creates a unique and memorable atmosphere, transforming each stay into an experience that guests will always carry in their memories. This combination of beauty and care results in a profound emotional connection that not only makes the experience timeless but also fosters brand loyalty. In this way, "luxury" becomes more than just a combination of services and products: it transforms into an emotional journey that leaves a lasting impression, making each stay a unique and unrepeatable experience.

In a nutshell, something that can only be lived in Borgo Egnazia... and Nowhere Else.

Source: The Authors thank Claudio Labartino, Sustainability & Digital Marketing Specialist @ Egnazia Group for the interview.

Luxury Restaurants and Fine Dining

Over the past twenty years, there has been a significant expansion of the luxury restaurant industry (Hwang and Hyun, 2013; Hyun and Kang, 2014; Velasco & Veflen, 2021). Restaurants have direct economic and environmental influences on food waste (e.g., Filimonau et al., 2023) and other environmental issues, such as plastic and packaging waste, GhG

[12] https://indigenus.it/

emissions, as well as water and energy consumption (Huang et al., 2023; Madanaguli et al., 2022). Moreover, luxury restaurants are often considered unsustainable because of their excess, utilization of rare materials (e.g., Beluga caviar), neglect of seasonality, and fat-rich plates (Batat & De Kerviler, 2020). At the same time, consumers are also becoming more aware of animal welfare issues, as for example the cruelty to ducks and geese to produce foie gras.[13] Through interviews with Michelin-starred chefs, Batat (2020) examined the pillars of sustainable food experiences in the luxury sector. The findings indicate that sustainable luxury food experiences are experiential offerings, that address both the tangible (functional) and intangible (emotional, social, symbolic) needs of customers by presenting healthy, creative, sensory, and enjoyable dishes in unique settings that feature exceptional service, while also advocating for ethical practices concerning the environment and the community (Batat, 2020). The converging themes represent five main pillars of a sustainable luxury food experience, categorized as intrinsic/internal elements (plate, pleasure, and place) and extrinsic/external factors (people and planet), which are synthetized in Table 5.2.

Moreover, the growing attention towards sustainability in this sector is also reflected by the institution of a new "award", namely the Michelin Green Star, which since 2021 is assigned to restaurants at the forefront of the industry when it comes to their sustainable practices (Michelin Guide, 2021). Inspectors consider ingredient sourcing (origin and seasonality), ecological footprint of the restaurant, management of waste, and resources (Michelin Guide, 2021). Because of their high reputation, fine-dining restaurants that get a Michelin Green Star could become important environmental leaders in the restaurant business and for the public in general (Huang et al., 2023).

People may have different ideas of how high-end restaurants should be, in terms of expensiveness, superior food quality, attentive service, pleasant atmosphere, and related feelings, like pleasure and luxury (Lee & Hwang, 2011). According to Chen et al. (2015, p. 238) luxury restaurant "is a full-service restaurant whose environment (e.g. décor, atmospherics and services) and products (e.g. food and beverages) are carefully

---

[13] https://www.peta.org/issues/animals-used-for-food/factory-farming/ducks-geese/foie-gras/

Table 5.2 Key principles of sustainability in luxury restaurants

Plate	• Pesticide free food ingredients • Respectful booking techniques • Reducing fat and sugar for healthy and creative cuisine • Offering natural cuisine • Reducing portions to prevent overeating • Meat free menus
Pleasure	• Enhancing food sensory and experiences • Pleasing guests while offering them healthy food • Enhancing food socialization • Food storytelling and authenticity
Place	• Adopting eco-friendly interior design • Using recycled material • Eliminating plastic • Adopting greener cleaning products • Reducing energy and light consumption
Planet	• Managing food waste • Adopting and promoting short circuits • Respecting seasonality • Considering animal welfare • Recycling
People	• Enhancing employee's well-being and training • Prompting food cultures and exchanges among communities • Connecting consumers, farmers, and food actors • Providing food and taste education to children • Helping small family businesses: local farmers, suppliers, and producers

Source: Adapted from Batat (2020)

prepared and presented, unique, superior in quality and conspicuous". This definition emphasizes the key role of aesthetics in crafting an immersive dining experience. The Italian chef Massimo Bottura (2019), owner of Osteria Francescana, suggests that the chef's duty is to uncover the intrinsic beauty of each ingredient and maximize its potential during every stage of its lifecycle. Furthermore, the chef deliberately exploits art as a source of inspiration, to elevate the creative process and provide a unique experience to his guests (https://osteriafrancescana.it/). One prominent example of the "artification" of the dining experience is Alinea restaurant in Chicago. Abrams (2013) highlights how the work of the chef Grant Achatz derives from a history of design in culinary innovation, which culminates in the scenic presentation of plates and relative aesthetic—and multisensorial—experiences. The restaurant interiors were designed to combine traditional architectural elements with contemporary design and materials (Gerzina & Sutton, 2016) while at the same time leaving the lead role to plates. Plated food plays on a diner's expectations, experience, and memory: It must be presented in a beautiful and original way, to provide a surprising, exciting, and memorable eating experience (Deroy et al., 2014). Some examples include the hanged bacon and the use of cutlery for unconventional food presentation proposed by the chef Achatz in his restaurant (Deroy et al., 2014).

Consumer perception in fine dining is deeply influenced by such aesthetic considerations, associating them with feelings of exclusivity and a heightened sensory journey. When a restaurant's design, ambiance, and presentation "synchronize" with the culinary artistry, it transforms a meal into an unforgettable event.

Beyond top restaurants, there's a growing tendency of luxury brands to expand their business towards gastronomy and hospitality, following the paradigmatic shift from tangible products to experiential economy (Kim et al., 2024). Some examples include spaces like the Dior Café in Miami, Le Café V by Louis Vuitton in Tokyo, the Louis Vuitton Restaurant in Saint Tropez, the Gucci Osteria in Florence and Beverly Hills (Motes, 2023), or even temporary cafes such as the Fendi takeover at Rinascente, in Milan (Floren, 2021) or the Dior pop-up store at the Restaurant and Beach Club Il Riccio in the island of Capri (Marelli, 2020). These branded spaces usually reflect the brand identity through both atmosphere and food aesthetics, enhancing loyalty to the restaurant and the parent brand, while at the same time being more accessible to a younger experience-driven customer base (Kim et al., 2024). Gucci Osteria by Massimo Bottura is a collection of Italian restaurants located worldwide, with varied identities, which still share a seasonal culinary philosophy inspired by originality, elegance, humor, and sensuality (https://www.gucciosteria.com/en/florence). To conclude, these new luxury experiences are designed to let consumers fully enter and live the brand's world, surrounded by visual elements such as brands, colors, and art pieces, carefully selected to reflect the brand's essence. Iconic and beautiful locations are part of the strategy, since they allow the creation of immersive experience surrounded by beauty and a sense of exclusivity. Whether in Michelin-starred restaurants, or in branded environments, the design of the experience will focus more and more on the principles of sustainability and inclusion, to signify the brand's commitment towards sustainable development.

Conclusion

Some emerging trends in luxury aesthetics include a shift towards more sustainable solutions. However, luxury firms should be aware of the potential damage of integrating sustainable design, since different consumers may perceive eco-friendly products as aesthetically unappealing (Han et al., 2017). However, we highlight that some Italian luxury brands already embraced "sustainable aesthetics" in a successful way. In Table 5.3 we summarize some successful examples.

Table 5.3 Major aesthetic trends and examples from Made in Italy

Aesthetic trend	Made in Italy example
Eco-friendly materials	**Salvatore Ferragamo** uses responsible materials such as organic cotton, viscose, and recycled and regenerated nylon, polyester, wool, and cashmere.[a] **Gucci Off the Grid** collection is made from recycled, organic, and responsibly sourced materials.[b]
Minimalism	Brands like **Bottega Veneta** and **Max Mara** are well known for their minimalist style, expressed through clear lines and light colors.
Bold colors and patterns	**Dolce & Gabbana** often propose baroque patterns and vivid colors inspired by Sicilian cultural heritage. Although characteristics of extravagant design, they reinforce the connection with the territory, thus encouraging the appreciation and preservation of unique, timeless designs.
Artisanal craftsmanship	**Brunello Cucinelli**[c] leverages its collaboration with a network of local artisans to foster creativity and innovation within the traditional cashmere sector.
Connection with nature	**Bulgari hotel**[d] in Rome exemplifies biophilic design via its urban jungle and incorporation of natural materials, merging luxury with a significant connection to nature.

Source: Authors' elaboration

[a]https://sustainability.ferragamo.com/en/products-materials/responsible-materials

[b]https://equilibrium.gucci.com/it/mindful-travel-with-gucci-off-the-grid/

[c]https://www.brunellocucinelli.com/company/the-value-of-italian-craftsmanship.html

[d]https://www.bulgarihotels.com/en_US/rome/the-hotel/green-initiatives

In the following chapter we will propose a new model of sustainable luxury consumption which is based on customer experiences with products and service environments, thus extremely related to individuals' aesthetic experiences.

Summary

Luxury brands can effectively promote sustainability without compromising their core values of excellent quality, exclusivity, and aesthetically pleasing experiences. At the same time, visual design is crucial for establishing and reinforcing a premium brand's identity, embodying the brand's fundamental essence, values, and market positioning. Indeed, luxury goods—as well as luxury service environments—are beautiful, refined, and aesthetically pleasing, which makes them very close to being considered a "*work of art*" (Venkatesh et al., 2010). Consequently, aesthetics significantly influences consumer decisions (Townsend, 2017; Townsend & Shu, 2010), through emotional responses, increased perceived value, uniqueness, and desirability of products or services (Holmqvist et al., 2020; Kapferer & Bastien, 2012).

The present chapter explored various industries within the luxury sector, focusing on personal luxury goods and luxury services. Categories such as fashion, jewelry, watches, and the automotive sector, alongside hospitality, spas, and fine dining, were examined. Each sector faces distinct challenges in transitioning toward sustainability, yet they are all adopting significant initiatives to minimize their environmental and social impact. Through notable examples, we observed how Italian brands are leading this shift, leveraging tradition, craftsmanship, and innovation to balance traditional values of luxury consumption with sustainable development. By blending aesthetics, sustainability, and functionality, these industries not only define contemporary luxury but also promote more responsible consumption.

References

Abrams, J. (2013). Mise en plate: The scenographic imagination and the contemporary restaurant. *Performance Research, 18*(3), 7–14.

Adıgüzel, F., & Donato, C. (2021). Proud to be sustainable: Upcycled versus recycled luxury products. *Journal of Business Research, 130*, 137–146.

Alfakhri, D., Harness, D., Nicholson, J., & Harness, T. (2018). The role of aesthetics and design in hotelscape: A phenomenological investigation of cosmopolitan consumers. *Journal of Business Research, 85*, 523–531.

Amatulli, C., De Angelis, M., Costabile, M. & Guido, G. (2017). Sustainable luxury brands: Evidence from research and implications for managers. Springer.

Amatulli, C., De Angelis, M., & Stoppani, A. (2021). The appeal of sustainability in luxury hospitality: An investigation on the role of perceived integrity. *Tourism Management, 83*, 104228.

Baker, J., Bentley, K., & Lamb, C., Jr. (2020). Service environment research opportunities. *Journal of Services Marketing, 34*(3), 335–346.

Batat, W. (2020). Pillars of sustainable food experiences in the luxury gastronomy sector: A qualitative exploration of Michelin-starred chefs' motivations. *Journal of Retailing and Consumer Services, 57*, 102255.

Batat, W., & De Kerviler, G. (2020). How can the art of living (art de vivre) make the French luxury industry unique and competitive? *Marché et organisations, 1*, 15–32.

Bellezza, S. (2023). Distance and alternative signals of status: A unifying framework. *Journal of Consumer Research, 50*(2), 322–342.

Bhandari, N., Garza-Reyes, J. A., Rocha-Lona, L., Kumar, A., Naz, F., & Joshi, R. (2022). Barriers to sustainable sourcing in the apparel and fashion luxury industry. *Sustainable Production and Consumption, 31*, 220–235.

Bitner, M. J. (1992). Servicescapes: The impact of physical surroundings on customers and employees. *Journal of marketing, 56*(2), 57–71.

Bottura, M. (2019). *Food for change* (Vol. 9). European Investment Bank.

Bovich, L. (2007). Renovation, conversion projects give renewed life to tired, aging hotels. *Hotel Business, 4*.

Cervellon, M. C., & Shammas, L. (2013). The value of sustainable luxury in mature markets: A customer-based approach. *Journal of Corporate Citizenship, 52*, 90–101.

Cesareo, L., Townsend, C., & Pavlov, E. (2023). Hideous but worth it: Distinctive ugliness as a signal of luxury. *Journal of the Academy of Marketing Science, 51*(3), 636–657.

Chan, J., Gao, Y., & McGinley, S. (2021). Updates in service standards in hotels: How COVID-19 changed operations. *International Journal of Contemporary Hospitality Management, 33*(5), 1668–1687.

Chattalas, M., & Shukla, P. (2015). Impact of value perceptions on luxury purchase intentions: A developed market comparison. *Luxury Research Journal, 1*(1), 40–57.

Chen, A., Peng, N., & Hung, K. P. (2015). The effects of luxury restaurant environments on diners' emotions and loyalty: Incorporating diner expectations into an extended Mehrabian-Russell model. *International Journal of Contemporary Hospitality Management, 27*(2), 236–260.

Cho, R. (2021). Why fashion needs to be more sustainable. State of the Planet. News from Columbia Climate School. Retrieved December 2024, from https://news.climate.columbia.edu/2021/06/10/why-fashion-needs-to-be-more-sustainable/

Chow Tai Fook, Ipsos. (2023). 2023 Jewellery Consumer Trends Report. Shaping the Future of Jewellery Consumption in China and Beyond: Preferences and Desires of Generation Z and Millennials. https://www.ctfjewellerygroup.com/file/Doc/pre231031_en.pdf

Dall'Olmo Riley, F., Pina, J. M., & Bravo, R. (2015). The role of perceived value in vertical brand extensions of luxury and premium brands. *Journal of Marketing Management, 31*(7–8), 881–913.

D'Arpizio, C., Levato, F., Steiner, A., de Montgolfier, J. (2024). *Long live luxury: Convergence to expand through turbolence.* Bain & Company. https://www.bain.com/insights/long-live-luxury-converge-to-expand-through-turbulence/

De Angelis, M., Amatulli, C., & Petralito, S. (2021). Luxury and sustainability: An experimental investigation concerning the diamond industry. In *Sustainable luxury and jewelry* (pp. 179–198). Springer Singapore.

Deloitte. (2023). Global Powers of Luxury Goods 2023. Retrieved August 2024, from https://www.deloitte.com/global/en/Industries/consumer/analysis/gx-cb-global-powers-of-luxury-goods.html

Deroy, O., Michel, C., Piqueras-Fiszman, B., & Spence, C. (2014). The plating manifesto (I): From decoration to creation. *Flavour, 3*, 1–11.

Díaz-Bustamante, M., Carcelén, S., & Puelles, M. (2015). The balance between the transmitted and the perceived image of luxury fragrance brands. *Journal of Modern Accounting and Auditing, 11*(10), 499–514.

Dinh, E., Mühlbacher, H., & Torchia, M. (2024). Necessary perceptions of family values and luxury characteristics for brand luxuriousness: Evidence from luxury watch brands. *Journal of Brand Management, 31*(1), 58–78.
Eckhardt, G. M., & Bardhi, F. (2020). New dynamics of social status and distinction. *Marketing Theory, 20*(1), 85–102.
Entwistle, J. (2023). *The fashioned body: Fashion, dress and modern social theory.* John Wiley & Sons.
Filimonau, V., Chiang, C. C., Wang, L. E., Muhialdin, B. J., & Ermolaev, V. A. (2023). Resourcefulness of chefs and food waste prevention in fine dining restaurants. *International Journal of Hospitality Management, 108*, 103368.
Floren, S. (2021). Il Fendi Caffe, in Rinascente. Retrieved September 2024, from https://www.lofficielitalia.com/food/fendi-caffe-bar-rinascente-milano-dove-foto-prezzi
Gerzina D., & Sutton, R. (2016). Inside Alinea 2.0: Look around the renovation and details on new dining options. Chicago Eater. https://chicago.eater.com/2016/5/3/11578144/alinea-photos-price-redesign-renovation#0
Godart, F., Claes, K., & Sgourev, S. V. (2020). Aesthetic and technological complexity in luxury watchmaking. In *Aesthetics and style in strategy* (pp. 49–76). Emerald Publishing Limited.
Grewal, D., Baker, J., Levy, M., & Voss, G. B. (2003). The effects of wait expectations and store atmosphere evaluations on patronage intentions in service-intensive retail stores. *Journal of Retailing, 79*(4), 259–268.
Han, J., Seo, Y., & Ko, E. (2017). Staging luxury experiences for understanding sustainable fashion consumption: A balance theory application. *Journal of Business Research, 74*, 162–167.
Hoffmann, S., & Akbar, P. (2023). *Consumer behavior: Understanding consumers – Designing marketing activities*. Springer Fachmedien Wiesbaden. https://doi.org/10.1007/978-3-65839476-9
Holmqvist, J., Ruiz, C. D., & Peñaloza, L. (2020). Moments of luxury: Hedonic escapism as a luxury experience. *Journal of Business Research, 116*, 503–513.
Hwang, J., & Hyun, S. S. (2013). The impact of nostalgia triggers on emotional responses and revisit intentions in luxury restaurants: The moderating role of hiatus. *International Journal of Hospitality Management, 33*, 250–262.
Hyun, S. S., & Kang, J. (2014). A better investment in luxury restaurants: Environmental or non-environmental cues?. *International Journal of Hospitality Management, 39*, 57–70.
Huang, Y., Hall, C. M., & Chen, N. (2023). The sustainability characteristics of Michelin green star restaurants. *Journal of Foodservice Business Research*, 1–26.

Kapferer, J. N. (2012). *The new strategic brand management: Advanced insights and strategic thinking*. Kogan Page Publishers.
Kapferer, J. N. (2017). The end of luxury as we knew it? *Advances in Luxury Brand Management*, 25–41.
Kapferer, J. N., & Michaut-Denizeau, A. (2015). Luxury and sustainability: A common future? The match depends on how consumers define luxury. *Luxury Research Journal, 1*(1), 3–17.
Kapferer, J. N., & Michaut-Denizeau, A. (2019). Are millennials really more sensitive to sustainable luxury? A cross-generational international comparison of sustainability consciousness when buying luxury. *Journal of Brand Management*, 1–13.
Karaosman, H., Perry, P., Brun, A., & Morales-Alonso, G. (2020). Behind the runway: Extending sustainability in luxury fashion supply chains. *Journal of Business Research, 117*, 652–663.
Kapferer, J. N. (2014). The artification of luxury: From artisans to artists. *Business Horizons, 57*(3), 371–380.
Kapferer, J. N., & Bastien, V. (2012). Luxe oblige. Editions Eyrolles.
Kim, D., Hyun, H., & Park, J. (2020). The effect of interior color on customers' aesthetic perception, emotion, and behavior in the luxury service. *Journal of Retailing and Consumer Services, 57*, 102252.
Kim, S., Cuevas, L., & Chang, H. J. (2024). Revisiting consumer loyalty: Luxury brand extension into restaurants and cafés. *International Journal of Retail & Distribution Management, 53*(2), 199–215. https://doi.org/10.1108/IJRDM-04-2024-0193
Kim, W. G., & Moon, Y. J. (2009). Customers' cognitive, emotional, and actionable response to the servicescape: A test of the moderating effect of the restaurant type. *International Journal of Hospitality Management, 28*(1), 144–156.
Kirillova, K., & Chan, J. (2018). "What is beautiful we book": Hotel visual appeal and expected service quality. *International Journal of Contemporary Hospitality Management, 30*(3), 1788–1807.
Kolling, C., Ribeiro, J. L. D., & de Medeiros, J. F. (2022). Performance of the cosmetics industry from the perspective of Corporate Social Responsibility and Design for Sustainability. *Sustainable Production and Consumption, 30*, 171–185.
Krishnan, C., Agarwal, N., & Gupta, S. (2022). Factors influencing consumer purchase intention of luxury branded perfumes. *Marketing i menedžment innovacij, 13*(3), 190–199.

Kumar, D. S., Purani, K., & Sahadev, S. (2017). Visual service scape aesthetics and consumer response: A holistic model. *Journal of Services Marketing, 31*(6), 556–573.
Lecointre-Erickson, D., Adil, S., Daucé, B., & Legohérel, P. (2024). The role of brick-and-mortar exterior atmospherics in post-COVID era shopping experience: A systematic review and agenda for future research. *Journal of Strategic Marketing, 32*(8), 1180–1194.
Lee, J. H., & Hwang, J. (2011). Luxury marketing: The influences of psychological and demographic characteristics on attitudes toward luxury restaurants. *International Journal of Hospitality Management, 30*(3), 658–669.
Leung, P. P., Wu, C. H., Ip, W. H., & Ho, G. T. (2019). Enhancing online-to-offline specific customer loyalty in beauty industry. *Enterprise Information Systems, 13*(3), 352–375.
Lim, W. M., & Endean, M. (2009). Elucidating the aesthetic and operational characteristics of UK boutique hotels. *International Journal of Contemporary Hospitality Management, 21*(1), 38–51.
Madanaguli, A., Dhir, A., Kaur, P., Srivastava, S., & Singh, G. (2022). Environmental sustainability in restaurants. A systematic review and future research agenda on restaurant adoption of green practices. *Scandinavian Journal of Hospitality and Tourism, 22*(4–5), 303–330.
Marelli, C. (2020. La dolce vita è in un negozio a picco sul mare di Capri, da raggiungere in barca. Elle Decor. Retrieved October 2024, from https://www.elledecor.com/it/lifestyle/a33273308/dioriviera-pop-up-store-dior-capri/
Michelin Guide. (2021). What is the Michelin Green Star? Retrieved January 2025, from https://guide.michelin.com/us/en/article/features/what-is-a-michelin-green-star
Miller, S. (2007). Fashion as art; is fashion art? *Fashion Theory, 11*(1), 25–40.
Motes, R. (2023). Luxury brands and their take on the restaurant industry. Retrieved July 2024, from https://www.society19.com/luxury-brands-and-their-take-on-the-restaurant-industry/
Munz, H. (2018). Crafting time, making luxury: The heritage system and artisan revival in the Swiss watch industry, 1975–2015. *Global Luxury: Organizational Change and Emerging Markets since the 1970s*, 197–218.
Norman, D. A. (2004). Emotional design: Why we love (or hate) everyday things.
Orth, U. R., & Wirtz, J. (2014). Consumer processing of interior service environments: The interplay among visual complexity, processing fluency, and attractiveness. *Journal of Service Research, 17*(3), 296–309.

Pijls, R., Galetzka, M., Groen, B. H., & Pruyn, A. T. (2023). Come in please: a virtual reality study on entrance design factors influencing the experience of hospitality. *Journal of Environmental Psychology, 90*, 102106.

Pijls, R., Groen, B. H., Galetzka, M., & Pruyn, A. T. (2017). Measuring the experience of hospitality: Scale development and validation. *International Journal of Hospitality Management, 67*, 125–133.

Pojani, E., Van Acker, V., & Pojani, D. (2018). Cars as a status symbol: Youth attitudes toward sustainable transport in a post-socialist city. *Transportation Research Part F: Traffic Psychology and Behaviour, 58*, 210–227.

Rafaeli, A., & Vilnai-Yavetz, I. (2004). Instrumentality, aesthetics and symbolism of physical artifacts as triggers of emotion. *Theoretical Issues in Ergonomics Science, 5*(1), 91–112. https://doi.org/10.1080/1463922031000086735

Rovai, S. (2016). *Luxury the Chinese way: The emergence of a new competitive scenario*. Springer.

Ruokamo, E., Räisänen, M., & Kauppi, S. (2022). Consumer preferences for recycled plastics: Observations from a citizen survey. *Journal of Cleaner Production, 379*, 134720.

Slater, D. (2002). Markets, materiality and the 'new economy'. *Market Relations and the Competitive Process*, 95–113.

Som, A., & Blanckaert, C. (2015). *The road to luxury: The evolution, markets, and strategies of luxury brand management*. John Wiley & Sons.

Sourvinou, A., & Filimonau, V. (2018). Planning for an environmental management programme in a luxury hotel and its perceived impact on staff: An exploratory case study. *Journal of Sustainable Tourism, 26*(4), 649–667.

Stankeviciute, R., & Hoffmann, J. (2011). The slippery slope of expansion-Look to the luxury sector to understand how brand extensions affect parent brands. *Marketing Management, 20*(4), 26.

Sukharevska, A. (2016). Promotion of luxury clothing brands in Ukraine. *European Cooperation, 3*(10), 124–132.

Sun, J. J., Bellezza, S., & Paharia, N. (2021). Buy less, buy luxury: Understanding and overcoming product durability neglect for sustainable consumption. *Journal of Marketing, 85*(3), 28–43.

Townsend, C. (2017). The price of beauty: Differential effects of design elements with and without cost implications in nonprofit donor solicitations. *Journal of Consumer Research, 44*(4), 794–815.

Townsend, C., & Shu, S. B. (2010). When and how aesthetics influences financial decisions. *Journal of Consumer Psychology, 20*(4), 452–458.

Velasco, C., & Veflen, N. (2021). Aesthetic plating and motivation in context. *International Journal of Gastronomy and Food Science, 24*, 100323.

Venkatesh, A., Joy, A., Sherry, J. F., Jr., & Deschenes, J. (2010). The aesthetics of luxury fashion, body and identify formation. *Journal of Consumer Psychology, 20*(4), 459–470.

Wang, Y. (2022). A conceptual framework of contemporary luxury consumption. *International Journal of Research in Marketing, 39*(3), 788–803.

West, A., & Hughes, J. (1991). An evaluation of hotel design practice. *Service Industries Journal, 11*(3), 362–380.

Williams, A. (2023). Sustainable sophistication: Luxury auto's eco-transformation. Nasdaq latest news. https://www.nasdaq.com/articles/sustainable-sophistication-luxury-autos-eco-transformation

Zanette, M. C., Pueschel, J., & Touzani, M. (2022). Re-arranging dressing practices: The role of objects in spreading ugly luxury. *Journal of Business Research, 145*, 784–800.

6

A New Perspective on Sustainable Luxury Consumption: Actionable Insights and Future Research

The final chapter of this book delves into recent advancements in sustainable luxury research, with a particular emphasis on how aesthetics, in terms of product and services spaces appearance, has the potential to trigger more sustainable behaviors. Despite its significant relevance, this area of study has been largely overlooked in previous research, except for some non-empirical work in the field of product design and green innovation. Throughout the previous chapters, we gradually built the case that luxury organizations, regardless of their sector, possess substantial resources and capabilities to take on the challenge of sustainable development. Furthermore, in Chap. 4 we presented examples of luxury and Made in Italy brands that have successfully embraced environmentally friendly practices, also through the effective leveraging of aesthetics in the process.

The present chapter presents a new perspective that bridges together insights from sustainable product design, consumer behavior, and neuroscience, going beyond the traditional view of the inherent sustainability of luxury that we discussed in Chap. 3. Indeed, previous research claiming that luxury is inherently sustainable grounds on a *product-based view* of sustainable luxury, suggesting that luxury products have a longer lifecycle essentially because they are of better quality. While we agree with

A. D'Aniello et al., *Aesthetics and Sustainable Luxury Consumption*, Palgrave Advances in Luxury, https://doi.org/10.1007/978-3-031-96833-4_6

this claim, we propose an innovative, *consumer-based view* of sustainable luxury, which focuses on consumers' willingness to use luxury products over a longer period of time driven by the experiences they live, over and beyond the concept of product quality. We do so by first bridging together luxury consumption values, sustainability, and aesthetic attributes in section "Bridging Sustainable Luxury Consumption Values and Aesthetics". Then, we briefly present some insights from recent unpublished empirical studies conducted by the authors of this book, as well as managerial implications and future research directions that might contribute to the development of the field in the future.

Bridging Sustainable Luxury Consumption Values and Aesthetics

Chapters 2 and 3 clarified that the evolution of luxury consumption values mirrors changes in societal priorities, advancements in technology, and demographic shifts. Through the analysis of consumer motivations as well as contemporary market behaviors, we explored both cognitive and emotional drivers of luxury buyers, showing that even if numerous academic studies consistently indicated that sustainability is not a pivotal consideration for most luxury consumers during their purchasing decisions (e.g., Kapferer & Michaut-Denizeau, 2014), we observe a progressive shift from the *conspicuousness* to the *conscientiousness* of consumption choices (e.g., Ki et al., 2017). Similarly, we extensively discussed the difference between *externalized* and *internalized* approaches to luxury consumption (Amatulli & Guido, 2012; Guido et al., 2020): while the former occurs when individuals purchase luxury mainly to gain social acceptance and are influenced by others' opinions and perceptions, the latter reflects consumers' desire to evoke positive emotions and express their unique taste and style (Amatulli & Guido, 2012). Consumers with internal motivations are more likely to choose sustainable luxury that aligns with their personal values (Hennigs et al., 2013), while externally motivated consumers place more emphasis on social recognition, making sustainability a secondary concern (Amatulli et al., 2017), unless it enhances their social image with relevant others (Griskevicius et al., 2010).

Before introducing the way in which consumer motives may also connect to different aesthetic attributes, at this stage it is crucial to emphasize that, given the complexity of human motivations (e.g., Moisander, 2007; Patall, 2012), none of these opposed constructs should be viewed as mutually exclusive. As we already pointed out in previous chapters, conspicuous consumption is driven by social comparison, and the desire to signal status, wealth, and success (Trigg, 2001; Veblen, 1899), thus it is often associated with unsustainable overconsumption and an externalized approach to luxury consumption (Amatulli & Guido, 2012). On the opposite side, conscientiousness deals with a desire to consider the impact of personal choices (Ki et al., 2017), thus it is more coherent with frugality, minimalism, responsibility, and the internalized approach to luxury consumption (Amatulli & Guido, 2012).

However, modern luxury is increasingly incorporating both dimensions, redefining what it means to consume luxuriously, because the expansion of luxury in multiple sectors and marketplaces has led to varied, innovative, and occasionally unforeseen modes of luxury consumption (Dubois et al., 2021). Consumers are increasingly pursuing benefits linked to luxury that transcend traditional luxury items; they may augment or substitute conventional luxury brands with alternative luxury or non-luxury products to bolster the status-signaling role of their possessions (Dubois et al., 2021).

In terms of consumers' motivation, the diffusion of luxury consumption requires looking for new ways to differentiate themselves from other consumers. Drawing on the acknowledgment that the main factor driving status signaling is distance from some reference categories,[1] an insightful conceptual work of Bellezza (2023) explicitly identified alternative signals of status based on six focal dimensions: *time, quantity, conspicuousness, aesthetics, culture, and pace of life*. Given that our core focus is the relationship between aesthetics and sustainable luxury consumption, we

[1] More specifically, according to Bellezza (2023) signals that are in some manner dissimilar to popular items, brands, and preferences will achieve the goal of dissociating the signaler from others.

will focus on those dimensions, described as emerging market trends, that have the potential to affect either aesthetic preferences or sustainable behaviors:

- **Vintage:** distance from mainstream signals is attained by returning to the past (i.e., the old) rather than projecting into the future (i.e., the new). Rather than introducing innovative, futuristic designs, numerous luxury brands carried out special initiatives and events or e-commerce platforms to commercialize refurbished vintage items (e.g. Gucci Vintage selection).[2] Due to the rapid introduction of new designs and the reduction of product life cycles, the time required for signals from the past to qualify as vintage may decrease over time: it may not take more than two decades for certain "hit" pieces of today to resurface as genuine vintage items in the future.
- **Voluntary simplicity, sustainability, and minimalism**: high-status consumers are decreasing their expenditures on material possessions to stand out from mainstream consumers. Frequently, this trend is driven by the desire to adopt a more sustainable lifestyle while avoiding abundance and constant desire for consumption which are typical of materialism (Wilson & Bellezza, 2022). The desire for sustainability may translate in sustainable actions such as choosing green products, owning luxury products for longer time, purchasing pre-loved items, consider renting over buying (e.g., Christodoulides et al., 2021; Turunen & Leipämaa-Leskinen, 2015), as well as favoring luxury experiences (e.g., sustainable tourism, wellness retreats, Michelin-starred dinners, etc.) over product possession (Atwal & Williams, 2017; D'Arpizio et al., 2021, 2024). The renowned Patagonia advertising campaign headlined "*Don't buy this jacket*", which was published in the New York Times in November 2011, serves as a compelling illustration of how brands can encourage people to purchase fewer items and eventually buy some with greater consciousness.[3]

[2] https://www.gucci.com/it/it/st/capsule/vault-vintage.

[3] https://eu.patagonia.com/it/it/stories/dont-buy-this-jacket-black-friday-and-the-new-york-times/story-18615.html.

- **Inconspicuous consumption and extravagant aesthetic choices**: one key attribute of conventional luxury items is their high degree of visibility and conspicuousness. Conspicuous markers, such as extravagant products, identifiable forms, or prominent logos, facilitate the process of signaling and the communication of desired identities. Those who consume luxury goods for personal effects, displaying an internalized approach to luxury consumption (Vigneron & Johnson, 1999; Amatulli & Guido, 2012), would be more attracted by inconspicuous and "silent" luxury products (Han et al., 2010). On the contrary, status seekers would be more likely to signal their social position through remarkable aesthetic features, which qualify what is called "loud" luxury products (Han et al., 2010) and the already mentioned "ugly" luxury which questions traditional standards of beauty (Sun et al., 2021). Differently from "loud" luxury products, the ugly ones allow luxury consumers to differentiate themselves; therefore, buying such items could be seen as a way to prevent imitation from others, satisfying their need for uniqueness (Bellezza, 2023).

While the constructs of conspicuousness and externalized consumption are intrinsically linked to visual display and aesthetics, the association between internalized approach, conscientious behavior, and aesthetics is less immediate, although extremely relevant and powerful. We propose that the relationship between consumer behavior and luxury aesthetics is dual (Fig. 6.1): in the pre-consumption phase, consumer motivation can help predict the aesthetic preferences of different customers; then, the aesthetics of a product (or services) influences actual (or inferred, Crolic et al., 2019) experiences and subsequent, post-consumption, sustainable behaviors. In the following section, we delve deeper into the connection between aesthetics and their impact on sustainable behaviors.

Fig. 6.1 The dual relationship between consumer behavior and aesthetics. (Source: Authors' elaboration)

Aesthetic Experiences: A Consumer-Based View of Sustainability in Luxury

Major research efforts about sustainable luxury have concentrated on a product-centric perspective that highlights the longevity of luxury goods owing to their exceptional quality. For example, Sun et al. (2021) prove that luxury itself can serve as a distinctive approach to sustainable consumption due to the exceptional durability of high-end products, which however is often neglected by customers at the moment of purchase. Following this argument, they suggest that luxury brands should emphasize product durability (defined as a product's ability to continue functioning over a long period of time under normal conditions of use without excessive expenditure on maintenance or repair; Cooper & Christer, 2016) in their communications to nudge consumers to buy high-end products.

Although we concur with this viewpoint, we contend that the durability of luxury products is heavily dependent on how long the customer will be willing to hold and use the product. Consequently, we advance a consumer-centric perspective on sustainable luxury, centered on the notion that users' experiences with a product influence its length of use (hereafter also LPU), surpassing issues of tangible durability and basic quality. Consequently, based on the importance of aesthetics in the luxury product experience (Cesareo et al., 2023), we propose that aesthetic attributes of luxury products can motivate consumers to use those products for longer, and thereby encourage sustainable consumption. Indeed, while previous research has established aesthetics as an important hedonic characteristic of luxury goods (e.g., Fuchs et al., 2013; Holmqvist et al., 2020; Kapferer & Bastien, 2009), whether luxury products' aesthetics can shape consumers' sustainable behaviors, and particularly LPU, remains unknown. However, recent research investigated the effect of specific aesthetics features such as color (Hagtvedt, 2020) and texture (Chen et al., 2025) on consumers' judgments of product durability: according to Hagtvedt (2020), darker-hued products are perceived as more durable yet less user-friendly compared to lighter-hued alternatives. Chen et al. (2025) found that consumers subconsciously associate rough textures with product longevity and strength and smooth textures with ease of use and comfort, testing this effect for different products (e.g., laptop, ear cuff, trolley). Furthermore, Wang et al. (2023) claimed that emotional durability, influenced by product aesthetics, serves as a catalyst for sustainable fashion consumption, and research in other fields has suggested that consumers might associate a product's aesthetics with its sustainability (e.g., Luchs et al., 2012). Wilson and Bellezza (2022) asserted that minimalistic consumption is intricately linked to sparse aesthetics, reflecting a desire for plain and unembellished designs. More generally, research has shown that consumers gain value from aesthetic stimuli across multiple domains (Schnurr & Wetzels, 2020; Hagtvedt & Patrick, 2008; Townsend, 2017; Yamamoto & Lambert, 1994). Our objective is to go beyond simple product evaluations by examining the actual responsible conduct of consumers as influenced by aesthetic experiences.

Previous non-empirical research suggests that the creation of durable items is substantially affected by aesthetics (Harper & Simonsen, 2017): the field of sustainable product design indicates some aesthetic attributes—such as simplicity, minimalism, logicality, utility, natural shapes, materials, upgradability, and modularity—that can enhance product durability (Zafarmand et al., 2003).

Recent unpublished qualitative research (De Angelis et al., 2025) has revealed five primary aspects of luxury aesthetics, comprising seventeen unique attributes (for a deeper explanation of the attributes see section "Aesthetics in Luxury"): "beauty" (encompassing visual impact, beauty, meticulousness, style, elegance), "recognizability" (comprising recognizability, color schemes, memorability, originality), "shapes" (encompassing minimalism, harmony, forms, and simplicity of lines), "craftsmanship" (comprising craftsmanship, design, and quality), and "cultural heritage" (consisting solely of cultural heritage). Beauty, recognizability, and shapes are sensory sensations stemming from a product's appearance; thus, we classify them as *outer aesthetics*. Craftsmanship and cultural heritage are fundamentally linked to the production of luxury goods; therefore, we classify them as *inner aesthetics*. Consistent with the literature on internalized versus externalized consumption approaches (Amatulli & Guido, 2012), we discovered that values related to workmanship and cultural heritage, representing inner aesthetics, were correlated with prolonged product usage. More specifically, across a series of experiments assessing both intended and actual luxury product usage (LPU, as measured in Sun et al., 2021), plus a field study based on automated text analysis on millions of user responses to luxury brands' social media content, we established that LPU is elevated when brands emphasize inner (as opposed to outer) aesthetics—and that this effect is due to the perceived continuity of luxury products. Indeed, inner aesthetics is able to foster a perception of timelessness and enduring quality, while outer aesthetics, often associated with status signaling and fashion trends, may result in shorter product usage cycles. Accordingly, the proposed effect is stronger among consumers with a reduced tendency towards conspicuous consumption (Chaudhuri et al., 2011, p. 217), as they are less influenced by external displays of luxury and more attuned to the intrinsic value of craftsmanship and heritage.

Even though aesthetic experiences are profoundly shaped by individual taste (Hoyer & Stokburger-Sauer, 2012; Wang, 2022), the field of empirical aesthetics has pinpointed some essential attributes that usually determine aesthetic preferences, which are: *contrast, clarity, color, unity, order, and symmetry* (Cesareo et al., 2023; Hagtvedt, 2022). Accordingly, in order to further reinforce our predictions, two additional research projects were aimed at exploring whether symmetry (D'Aniello et al., 2024) could affect sustainable behaviors, still in terms of expected and actual LPU. Symmetry is defined as the simplest case of balance, represented by mirror-inverted halves, is readily discernible and linked to simplicity, timelessness, and continuity (Treder, 2010; Pombo & Velasco, 2021). Non-complex, symmetrical aesthetics are easily detectable (Treder, 2010) and thus are an effective means of activating recipients' immediate gratification, that is, their anticipated experiences when interacting with an object (e.g., how it should feel when handled, held, lifted, or worn) (Harper & Simonsen, 2017). Based on these arguments, in our recent work based on results from experimental studies, participants were exposed to controlled visual stimuli depicting either symmetric or asymmetric luxury products. As a result, we consistently found that symmetric (vs. asymmetric) product design increased both intended LPU for fictitious luxury products and actual LPU for real luxury items owned by respondents. Given that product durability depends on both functional and stylistic benefits (Levinthal & Purohit, 1989), we identified stylistic continuity as a potential underlying driver of this effect. Again, it is stylistic continuity—defined as the degree to which a product endures through time and trends (derived from the continuity dimension of the brand authenticity; Morhart et al., 2015)—that might explain this effect. We specifically leverage the notion that symmetry imparts visual simplicity and reduces the perception of brands as trendy or up-to-date (Bajaj & Bond, 2018, p. 402), so that customers would regard symmetric luxury products as more timeless compared to asymmetric ones. In other words, stylistic benefits are just as important as functional benefits in prolonging the lifespan of luxury fashion products.

The third last empirical study that we present in this section focuses on harmony of contrasts. Several studies have found that consumers generally prefer harmony to disharmony, even in a composition with a variety of elements (Locher et al., 1999; Lomas & Xue, 2022). In the consumption realm using a variety of products, Veryzer Jr and Hutchinson (1998) found that as the harmony (used interchangeably with unity) in the design of products increased, there was a corresponding non-linear super-additive increase in aesthetic preference for the product. According to the gestalt tradition of cognitive psychology, humans tend to see things that are close together or that look, sound, or feel the same as belonging together (Sharps & Wertheimer, 2000). High levels of perceived harmony in the design of products make them inherently attractive (Coates, 2003).

Based on these investigations and in line with the argument, Schloss and Palmer (2011) defined the harmony of contrast as follows:

- harmony of contrast of scale for colors that differ significantly in lightness and are the same in hue,
- harmony of contrast of hues for colors that differ in lightness and are similar in hue,
- harmony of contrast of colors for colors that are different in hue and different in lightness (although the lightness difference is claimed to be auxiliary).

Due to the interest in harmony and contrast in product aesthetics, an exploratory study was conducted comparing two products (t-shirts): one with contrasting colors and the other without contrast. The stimuli used in the study consisted of two t-shirts. One was considered with contrast (with blue, white, and red colors) and the other was no-contrast (plain white).

The results confirmed that the two t-shirts were perceived as significantly different in terms of contrast. As before, the dependent variable was LPU and results showed higher LPU for the no-contrast stimulus as per our hypothesis.

Additionally, we introduced other variables to examine potential mediators in the contrast-LPU relationship. These variables were: attractiveness, unisex appeal, perceptual fluency, sustainability, trendiness, atypicality, and authenticity. After conducting several analyses, we found that the mediating variable in this effect was authenticity. Specifically, the effect can be described as follows: With-contrast = lower perceived authenticity = lower product durability.

The study provides valuable insights into the intersection of product aesthetics and consumer perception, specifically regarding contrast and sustainability. Our findings demonstrate that authenticity plays a key mediating role in the relationship between contrast and product durability, suggesting that visual elements in product design can significantly influence consumer behavior and sustainability metrics. This opens avenues for future research on authenticity as a psychological mediator in design aesthetics and its broader impact on sustainable consumer practices.

To the best of authors' knowledge, these empirical studies are the first to test the differential effects of luxury aesthetics dimensions on sustainable choices.

The Luxury Aesthetics & Sustainability Framework

Based on the above discussion, we introduce a novel framework on Sustainable Luxury Consumption (Fig. 6.2) which establishes a crucial link between luxury, aesthetics, and sustainable choices. This new model may serve as a basis for managerial decisions, as well as a tool for stimulating research in the field.

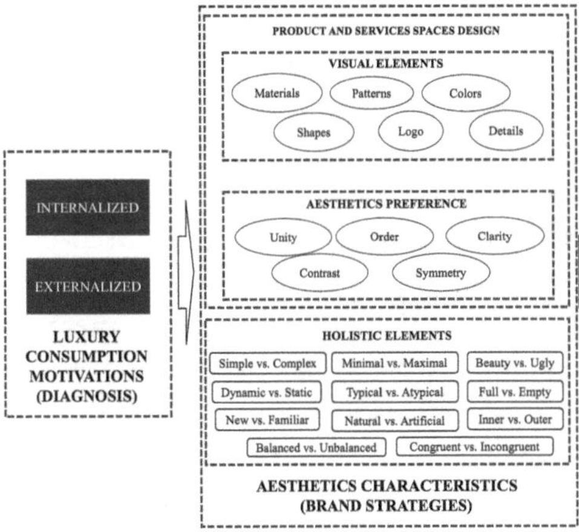

Fig. 6.2 The aesthetics and sustainable luxury consumption framework. (Source: Authors' elaboration)

As discussed in section "Aesthetic Experiences: A Consumer-Based View of Sustainability in Luxury" the relationship between consumer behavior and aesthetics is dual: first, consumer motivation—which we classify into internalized and externalized categories (Amatulli & Guido, 2012) —can serve as a predictive tool for anticipating the type of aesthetic attributes consumers find appealing, as outlined in the initial phase of our model. Secondly, aesthetics itself plays a pivotal role as a driver of consumer experiences, which, in turn, has profound implications for sustainable consumption behaviors. Specifically, aesthetic appreciation fosters emotional attachment and perceived product value, leading to behaviors such as extended product usage (LPU), mindful ownership, and reduced disposability—all of which are crucial elements in sustainable luxury consumption. This interplay is central to the second phase of our model, where we emphasize how aesthetics can influence long-term sustainability-oriented decisions by shaping the consumer's emotional and sensory engagement with luxury goods. Indeed, aesthetic experience is a cognitive process wherein imagination, cognition, and perception converge to foster emotional involvement with products or services, hence generating memorable experiences (Wang, 2024). At the same time, neuroscientific research indicates that aesthetically pleasant stimuli can activate brain reward systems, subsequently eliciting a favorable reaction from the consumer (e.g., Yun et al., 2022, see Chap. 4 for further details). In line with the work of Yun et al. (2022), we distinguish between physiological and psychological mechanisms: the former pertains to the brain's response (e.g., brain reward system), whilst the latter emphasizes that aesthetic preferences are influenced by identity, taste, and self-perception through cognitive processes. In terms of brain response, throughout Chap. 4 we introduced key concepts of Neuroaesthetics, an emerging discipline that combines neuroscience and aesthetics to understand how the human brain processes visual and emotional stimuli. Indeed, it is known that the activation of certain areas of the brain, such as the reward system, plays a crucial role in the appreciation of beauty in objects and other visual stimuli, thus also luxury. To illustrate, Fig. 6.3 represents different brain regions that have been related to aesthetics and the perception of beauty. Furthermore, the figure highlights the "default mode network" (DMN; yellow for the inferior parietal lobule; pink for the medial prefrontal cortex and blue for the posterior cingulate cortex) previously associated with self-referential mentation. Indeed, the DMN supports the integration of emotional and cognitive processes, affecting the overall experience of beauty and originality: within these networks, it is possible to observe sensitivity to positive and negative emotional aspects of aesthetic experience, and evidence for individual differences correlated with personal differences in aesthetic evaluation (Vessel et al., 2012). Accordingly, we support the idea that integrating neuroscientific insights with experimental and behavioral research could provide the most

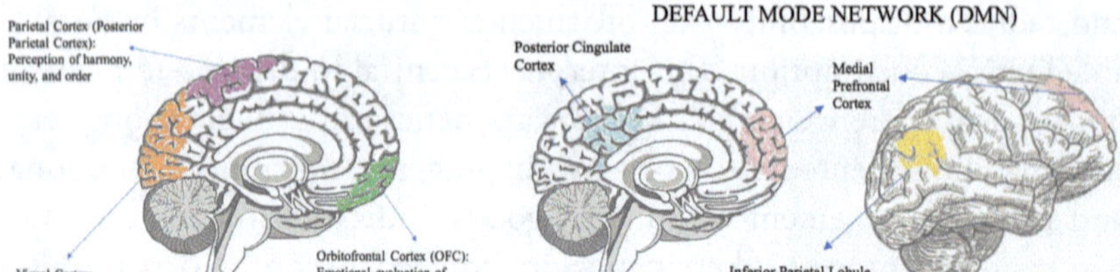

Fig. 6.3 Brain regions activated through aesthetics. (Source: Authors elaboration)

comprehensive understanding of consumers' experiences with the aesthetic stimuli of luxury products and services.

Aesthetics then takes center stage, acting as a bridge between consumers' motivations and experiences. Based on insights from extant literature (e.g., Amatulli et al., 2021; Cesareo et al., 2023; Yun et al., 2022), we classify aesthetic dimensions distinguishing between visual elements (such as color, shape, materials, logo) and holistic elements, which encompass both objective characteristics (i.e. complexity, atypicality) and subjective evaluations (e.g., beauty vs. ugly, trendiness, etc.). At the same time, we highlight some features which have been identified as elements that drive aesthetic preference, namely contrast, clarity, unity, order, and symmetry (Hagtvedt, 2022; Cesareo et al., 2023). On one side, aesthetics as preferences pertains to tangible aspects such as product and services spaces design, since these elements are those that could be leveraged by brands when defining their own strategies. On the other hand, aesthetics as stimuli opens the door to consumer behavior and neuroscience, exploring how visual elements such as contrast and symmetry elicit emotional and cognitive responses (e.g., perceived authenticity or stylistic continuity, as tested in our empirical studies, see section "Aesthetic Experiences: A Consumer-Based View of Sustainability in Luxury") within the "black box" of the consumer.

The proposed model serves as a comprehensive framework for understanding and leveraging the interplay of luxury, aesthetics, and sustainability from both academic and managerial perspectives.

Managerial Implication and Future Research Direction

Luxury aesthetics constitutes a multifaceted construct characterized by different dimensions that may have a differential effect on product usage over time. From a managerial perspective, our model provides a strategic framework for designing and positioning luxury products that resonate with contemporary consumer expectations. The incorporation of LPU as a key outcome that creates value underscores the importance of creating durable products that enhance perceived quality while aligning with sustainability principles. By merging durability with meticulously crafted aesthetics, brands can create offerings that appeal both emotionally and functionally to their target audiences.

Key components such as symmetry, contrast, and the "Made in" concept reinforce the perception of authenticity, continuity, and exclusivity. These elements, when combined with a profound understanding of consumer behavior, empower luxury brands to merge traditional notions of exclusivity with the growing demand for sustainability, authenticity, and social responsibility, thus contributing to effective luxury brand management.

For example, based on the results of our empirical studies to promote product continuity and sustainable behaviors, luxury brands should emphasize dimensions of inner aesthetics, communicating the craftsmanship and cultural heritage of luxury products in their marketing communication if they want to nudge consumers to use such products longer. Indeed, inner aesthetics not only aligns with sustainability goals but also redefines consumer engagement with luxury goods. Analogously, symmetry—an apparently simple design attribute—is a subtle yet effective cue for influencing green behaviors. Luxury brand managers who want to promote the durability of their products should prioritize conventional standards of beauty (i.e., favor symmetric shapes), which consumers perceive as more timeless and worthy of being retained for longer. Moreover, even if product durability is not always a prominent factor when making choices (Sun et al., 2021), consumers are usually confident that luxury purchases will be functionally durable. Therefore, communication about product durability should emphasize their permanent stylistic benefits, in terms of product continuity. Most of all, symmetry can be an effective tool for encouraging consumers to use luxury fashion products for longer. This issue is particularly relevant considering wealthy consumers' growing interest in the so-called "ugly luxury": visually unappealing products that are mainly chosen as status signals (e.g., Bellezza, 2023; Cesareo et al., 2023). Moreover, it is important for luxury brands to align design elements, such as color contrast, with consumer perceptions of authenticity to enhance product longevity and sustainability. Specifically, designing products with lower contrast may bolster perceptions of authenticity, thereby encouraging longer product use and reducing environmental impact. Brands aiming to position themselves as sustainable can leverage these findings to create new product designs that resonate with consumer values while also fostering sustainability goals.

A key challenge for fashion designers going forward will be managing the delicate balance between innovation and creativity on the one hand and long-term sustainability on the other.

The latter represents fertile ground for future research, since aesthetics for sustainable luxury is still overlooked and requires more effort to build a deeper understanding of this phenomenon. In this sense, our model contributes to research in the field by providing a solid base for future studies, identifying key aesthetic elements that could be explored in the future.

As for example, our findings are limited to specific aesthetic dimensions (e.g., inner vs. outer; symmetry vs. asymmetry; no-contrast vs. with contrast) but there are many other dimensions, such as color, unity, order, design principles, materials, graphics, that could positively impact experiences and sustainable consumption (e.g., Cesareo et al., 2023). Relatedly, most products combine two or more aesthetic features, which can simultaneously convey different aesthetic experiences. Therefore, future research should explore whether other aesthetic features, or their combination, produce differential effects on LPU.

Furthermore, future research could investigate additional processes such as consumers' emotional responses (e.g., Kumar & Garg, 2010), perceptions of functional obsolescence (Levinthal & Purohit, 1989), or processing fluency, in order to explore the underlying mechanisms that explain these effects. For example, it could be that asymmetric products are perceived as more perishable because of their unusual shape.

Third, it is important to explore variances among different product categories or sectors, as well as the impact of previous exposure to a similar design in that category (e.g., De Angelis et al., 2017). For example, it could be interesting to test whether the effect we uncovered is even stronger for fast-fashion products (e.g., Sun et al., 2021).

Finally, future research should consider collecting and analyzing field and neuroscientific data about the aesthetic features of fashion items that consumers retain in their closets for longer. By integrating neuroscience, our theoretical model establishes a direct link between aesthetic perceptions and purchase intentions, offering a robust theoretical foundation for further research. Such work could provide actionable insights and stimulate debate on green product development.

As sustainability becomes more and more relevant, being one of the core attributes to which consumers attach value especially in more mature markets (Osburg et al., 2024), it may even be that some consumers will choose to acquire high-end products and services because they acknowledge the inherent sustainability of luxury (vs. mass-market products).

Conclusions

The influence of aesthetics on sustainable luxury consumption represents a compelling subject that offers numerous avenues for further investigation. Considering the significance of aesthetics in the consumption domain, we propose that aesthetic attributes of luxury products, packaging, retail spaces, and service environments can influence individuals' perceptions and behaviors towards sustainable actions either consciously or unconsciously, through various cognitive and emotional processes. As sustainability becomes mainstream for the luxury market, and the damages of climate change are no longer questionable, a bigger effort from firms is required and the exceptional quality of luxury productions and services may no longer be enough. Although our work is not free of limitations, we genuinely aspire for it to provide a foundation for stimulating academic debate and managerial reflections for a sustainable future: one in which aesthetics act as a catalyst for consumer education and luxury establishes a standard for responsible enterprises.

Summary

The final chapter of this book proposes a new model of sustainable luxury, which emphasizes the critical role of aesthetics in shaping consumer behavior and fostering more responsible consumption. More specifically, we go beyond the traditional product-based view of sustainable luxury, centered around high quality and craftsmanship that guarantee extended functional durability (e.g. Sun et al., 2021), by introducing a consumer-centric approach, that considers not only the inherent sustainability of luxury products and services, but also how aesthetic attributes influence consumer experiences and subsequent sustainable behaviors.

At the beginning of the chapter, we briefly highlighted once again that luxury consumption is undergoing a transformation, reflecting broader societal shifts: from traditional conspicuous consumption to alternative status signals (Bellezza, 2023). For instance, some may opt for understated, timeless designs (often called "quiet luxury") as a means of demonstrating sophistication and environmental awareness, while others may choose the same products just because they are relevant for significant others. Then, we delve deeply into the interplay between aesthetics and sustainability, proposing that aesthetic attributes of luxury products and services shape consumer experiences, which in turn affect post-purchase behaviors, such as product longevity. Minimalist designs and natural colors to symmetry and craftsmanship are shown to play a significant role in fostering emotional attachment and encouraging longer product use. For example, symmetry is associated with perceptions of timelessness and durability, while no-contrast designs are seen as more authentic, enhancing their appeal over time.

Despite the advancements, significant gaps remain in the study of sustainable luxury. For instance, while prior research has focused on the quality and durability of luxury products, less attention has been given to the role of consumer experience in determining product longevity. Our proposed consumer-based perspective and the novel theoretical framework on aesthetics and sustainable luxury consumption address this gap by examining how aesthetic elements influence product use and sustainability behaviors, paving the way for a deeper understanding of this dynamic.

In conclusion, through a deeper understanding and awareness of the potential impact of aesthetics on consumer choices, luxury brands may not only enhance their appeal but also significantly contribute to a better world.

References

Amatulli, C., De Angelis, M., Costabile, M., & Guido, G. (2017). *Sustainable luxury brands: Evidence from research and implications for managers*. Springer.

Amatulli, C., De Angelis, M., & Donato, C. (2021). The atypicality of sustainable luxury products. *Psychology & Marketing, 38*(11), 1990–2005.

Amatulli, C., & Guido, G. (2012). Externalised vs. internalised consumption of luxury goods: Propositions and implications for luxury retail marketing. *The International Review of Retail, Distribution and Consumer Research, 22*(2), 189–207.

Atwal, G., & Williams, A. (2017). Luxury brand marketing – The experience is everything! *Advances in Luxury Brand Management*, 43–57.

Bajaj, A., & Bond, S. D. (2018). Beyond beauty: Design symmetry and brand personality. *Journal of Consumer Psychology, 28*(1), 77–98.

Bellezza, S. (2023). Distance and alternative signals of status: A unifying framework. *Journal of Consumer Research, 50*(2), 322–342.

Cesareo, L., Townsend, C., & Pavlov, E. (2023). Hideous but worth it: Distinctive ugliness as a signal of luxury. *Journal of the Academy of Marketing Science, 51*(3), 636–657.

Chaudhuri, H.R., Mazumdar, S., & Ghoshal, A., (2011). Conspicuous consumption orientation: Conceptualisation, scale development and validation. *Journal of Consumer Behaviour, 10*(4), 216–224.

Chen, J., Wu, P., Jiang, Y., & Lv, L. (2025). Rough is durable, smooth is user-friendly: Effects of surface visual roughness on consumers' perception and product preference. *Journal of Retailing and Consumer Services, 84*, 104196.

Christodoulides, G., Athwal, N., Boukis, A., & Semaan, R. W. (2021). New forms of luxury consumption in the sharing economy. *Journal of Business Research, 137*, 89–99.

Coates, D. (2003) Watches tell more than time: Product Design, Information, and the Quest for Elegance McGraw-Hill, New York (2003)

Cooper, T., & Christer, K. (2016). Marketing durability. In *Longer lasting products* (pp. 273–296). Routledge.

Crolic, C., Zheng, Y., Hoegg, J., & Alba, J. W. (2019). The influence of product aesthetics on consumer inference making. *Journal of the Association for Consumer Research, 4*(4), 398–408.

D'Aniello, A., Amatulli, C., De Angelis, M., & Pozharliev, R. (2024). Aesthetics for longevity: How symmetric design enhances sustainable luxury consumption. In SIM Conference 2024 conference proceedings.

D'Arpizio, C., Levato, F., Prete, F., Gault, C., & de Montgolfier, J. (2021). The Future of Luxury: bouncing back from Covid-19. Available at https://www.bain.com/insights/the-future-of-luxury-bouncing-back-from-covid-19/

De Angelis, M., Adıgüzel, F., & Amatulli, C. (2017). The role of design similarity in consumers' evaluation of new green products: An investigation of luxury fashion brands. *Journal of Cleaner Production, 141*, 1515–1527.

De Angelis, M., Amatulli, C., Pozharliev, R., Cardamone, E., & Di Cioccio, M. (2025). How Luxury Aesthetics Shapes Sustainable Consumption, working paper.

Dubois, D., Jung, S., & Ordabayeva, N. (2021). The psychology of luxury consumption. *Current Opinion in Psychology, 39*, 82–87.

Griskevicius, V., Tybur, J. M., & Van den Bergh, B. (2010). Going green to be seen: Status, reputation, and conspicuous conservation. *Journal of Personality and Social Psychology, 98*(3), 392.

Guido, G., Amatulli, C., Peluso, A. M., De Matteis, C., Piper, L., & Pino, G. (2020). Measuring internalized versus externalized luxury consumption motivations and consumers' segmentation. *Italian Journal of Marketing, 2020*, 25–47.

Hagtvedt, H. (2020). Dark is durable, light is user-friendly: The impact of color lightness on two product attribute judgments. *Psychology & Marketing, 37*(7), 864–875. https://doi.org/10.1002/mar.21268

Hagtvedt, H. (2022). A brand (new) experience: Art, aesthetics, and sensory effects. *Journal of the Academy of Marketing Science, 50*(3), 425–428.

Hagtvedt, H., & Patrick, V. M. (2008). Art and the brand: The role of visual art in enhancing brand extendibility. *Journal of Consumer Psychology, 18*(3), 212–222.

Han, Y. J., Nunes, J. C. and Drèze, X. (2010). Signaling status with luxury goods: The role of brand prominence. *Journal of Marketing*, 74(4), 15–30.

Harper, K., & Simonsen, R. R. (2017). *Aesthetic sustainability: Product design and sustainable usage*. Routledge.

Hennigs, N., Wiedmann, K. P., Klarmann, C., & Behrens, S. (2013). Sustainability as part of the luxury essence: Delivering value through social and environmental excellence. *Journal of Corporate Citizenship, 52*, 25–35.

Holmqvist, J., Ruiz, C. D., & Peñaloza, L. (2020). Moments of luxury: Hedonic escapism as a luxury experience. *Journal of Business Research, 116*, 503–513.

Hoyer, W. D., & Stokburger-Sauer, N. E. (2012). The role of aesthetic taste in consumer behavior. *Journal of the Academy of Marketing Science, 40*(1), 167–180.

Kapferer, J. N., & Michaut-Denizeau, A. (2014). Is luxury compatible with sustainability? Luxury consumers' viewpoint. *Journal of Brand Management, 21*, 1–22.

Ki, C., Lee, K., & Kim, Y. K. (2017). Pleasure and guilt: How do they interplay in luxury consumption? *European Journal of Marketing, 51*(4), 722–747.

Kumar, M., & Garg, N. (2010). Aesthetic principles and cognitive emotion appraisals: How much of the beauty lies in the eye of the beholder?. *Journal of Consumer Psychology, 20*(4), 485–494.

Levinthal, D. A., & Purohit, D. (1989). Durable goods and product obsolescence. *Marketing Science, 8*(1), 35–56.

Locher, P. J., Stappers, P. J., & Overbeeke, K. (1999). An empirical evaluation of the visual rightness theory of pictorial composition. *Acta psychologica, 103*(3), 261–280.

Lomas, J. D., & Xue, H. (2022). Harmony in design: A synthesis of literature from classical philosophy, the sciences, economics, and design. *She Ji: The Journal of Design, Economics, and Innovation, 8*(1), 5–64.

Luchs, M. G., Brower, J., & Chitturi, R. (2012). Product choice and the importance of aesthetic design given the emotion-laden trade-off between sustainability and functional performance. *Journal of Product Innovation Management, 29*(6), 903–916.

Moisander, J. (2007). Motivational complexity of green consumerism. *International Journal of Consumer Studies, 31*(4), 404–409.

Morhart, F., Malär, L., Guèvremont, A., Girardin, F., & Grohmann, B. (2015). Brand authenticity: An integrative framework and measurement scale. *Journal of Consumer Psychology, 25*(2), 200–218.

Osburg, V. S., Yoganathan, V., Bartsch, F., Diallo, M. F., & Liu, H. (2024). How sustainable luxury influences product value perceptions and behavioral intentions: A comparative study of emerging vs. developed markets. *Journal of Business Ethics, 191*(4), 713–738.

Patall, E. A. (2012). The motivational complexity of choosing: A review of theory and research. In M. R. Ryan (Ed.), *The Oxford handbook of human motivation* (pp. 248–279). Oxford University Press. https://doi.org/10.1093/oxfordhb/9780195399820.013.0015

Pombo, M., & Velasco, C. (2021). How aesthetic features convey the concept of brand premiumness. *Psychology & Marketing, 38*(9), 1475–1497.

Schloss, K. B., & Palmer, S. E. (2011). Aesthetic response to color combinations: Preference, harmony, and similarity. *Attention, Perception, & Psychophysics, 73*, 551–571.

Schnurr, B., & Wetzels, M. (2020). The conditional enjoyment-enhancing effect of shipping box aesthetics. *Journal of Retailing and Consumer Services, 57*, 102213.

Sharps, M. J., & Wertheimer, M. (2000). Gestalt Perspectives on Cognitive Science and on Experimental Psychology. *Review of General Psychology, 4*(4), 315–336.

Sun, J. J., Bellezza, S., & Paharia, N. (2021). Buy less, buy luxury: Understanding and overcoming product durability neglect for sustainable consumption. *Journal of Marketing, 85*(3), 28–43.

Townsend, C. (2017). The price of beauty: Differential effects of design elements with and without cost implications in nonprofit donor solicitations. *Journal of Consumer Research, 44*(4), 794–815.

Treder, M. S. (2010). Behind the looking-glass: A review on human symmetry perception. *Symmetry, 2*(3), 1510–1543.

Trigg, A. B. (2001). Veblen, Bourdieu, and conspicuous consumption. *Journal of Economic Issues, 35*(1), 99–115.

Turunen, L. L. M., & Leipämaa-Leskinen, H. (2015). Pre-loved luxury: Identifying the meanings of second-hand luxury possessions. *Journal of Product & Brand Management, 24*(1), 57–65.

Veblen, T. (1899). *The theory of the leisure class: An economic study of institutions.* Random House.

Veryzer, R. W., Jr., & Hutchinson, J. W. (1998). The influence of unity and prototypicality on aesthetic responses to new product designs. *Journal of Consumer Research, 24*(4), 374–394.

Vessel, E. A., Starr, G. G., & Rubin, N. (2012). The brain on art: Intense aesthetic experience activates the default mode network. *Frontiers in Human Neuroscience, 6*, 66.

Vigneron, F., & Johnson, L. W. (1999). A review and a conceptual framework of prestige-seeking consumer behavior. *Academy of Marketing Science Review, 1*(1), 1–15.

Wang, Y. (2022). A conceptual framework of contemporary luxury consumption. *International Journal of Research in Marketing, 39*(3), 788–803.

Wang, R. (2024). A Review of the Impact of Aesthetic Experience on Consumers' Purchase Decisions. *Highlights in Business, Economics and Management, 27*, 312–318.

Wang, M., Kim, E., & Du, B. (2022). Promoting emotional durability and sustainable fashion consumption through art derivatives design methods. *The Design Journal, 25*(5), 789–806.

Wang, Y., Xu, A. J., & Zhang, Y. (2023). L'Art Pour l'Art: Experiencing art reduces the desire for luxury goods. *Journal of Consumer Research, 49*(5), 786–810.

Wilson, A. V., & Bellezza, S. (2022). Consumer minimalism. *Journal of Consumer Research, 48*(5), 796–816.

Yamamoto, M., & Lambert, D. R. (1994). The impact of product aesthetics on the evaluation of industrial products. *Journal of Product Innovation Management, 11*(4), 309–324.

Yun, H., Xiao, L., Jingjing, L., & Lin, W. (2022). Visual elements in brand aesthetics: A review and prospects. *Foreign Economics & Management, 44*(08), 121–137. https://doi.org/10.16538/j.cnki.fem.20220513.301

Zafarmand, S. J., Sugiyama, K., & Watanabe, M. (2003). Aesthetic and sustainability: The aesthetic attributes promoting product sustainability. *The Journal of Sustainable Product Design, 3*(3), 173–186.

Index

A. D'Aniello et al., *Aesthetics and Sustainable Luxury Consumption*, Palgrave Advances in Luxury, https://doi.org/10.1007/978-3-031-96833-4

Kolibri 360 GmbH
Ferdinand-Jühlke-Straße 7
99095 Erfurt, Deutschland
produktsicherheit@kolibri360.de